Cereals and Cereal Products

Third Supplement to McCance and Widdowson's

The Composition of Foods

i

Cereals and Cereal Products

Third Supplement to

McCance and Widdowson's

The Composition of Foods

B. Holland, I. D. Unwin and D. H. Buss

The Royal Society of Chemistry
and
Ministry of Agriculture, Fisheries and Food

The Royal Society of Chemistry
The University
Nottingham NG7 2RD
UK

Tel.: (0602) 507411 Telex: 37488

ISBN 0-85186-743-X

Orders should be addressed to:
The Royal Society of Chemistry
Distribution Centre
Letchworth
Herts. SG6 1HN
UK

Xerox Ventura Publisher™ output photocomposed by
Spectrum Graphics, Basford, Nottingham

Printed in the United Kingdom by
Unwin Brothers Limited, Old Woking, Surrey

PREFACE

Following publication of the fourth edition of McCance and Widdowson's *The Composition of Foods* in 1978, the Ministry of Agriculture, Fisheries and Food took on the responsibility for maintaining and updating the official tables of food composition in the United Kingdom. In 1987 the Ministry joined with the Royal Society of Chemistry to begin production of a computerised UK National Nutrient Databank, from which this and future supplements to *The Composition of Foods* will be produced, as will an increasing variety of machine-readable products. This method of production will give greater accuracy, consistency and flexibility than is possible with manual compilation.

The availability of products from, and access to, the UK National Nutrient Databank will be developed simultaneously with the printed tables. As the nutrient values for each food group are updated and other data added, these will be merged with previously published data (principally from the fourth edition of *The Composition of Foods*) to bring the Databank progressively up-to-date. Thus nutrient values for the whole range of foods will be available for use with computer software, but only as material is revised will supporting information such as source references be added to the Databank. Some forms of access, for example through online search systems, may only become viable as the revision of food groups nears completion. The comments of users and potential users will influence the timing as well as the content and appearance of the products.

Epidemiological studies involving more than one country have heightened interest in international collaboration on the content of food composition tables. Much work has been done on food coding which seeks to describe food in detail and facilitate more precise computer retrieval, as well as to improve the comparability of nutrient data within and between countries. The RSC and MAFF will monitor and, where appropriate, participate in such developments so that the benefits can be available to users of the UK National Nutrient Databank.

CONTENTS

ACKNOWLEDGEMENTS

A large number of people have helped in the preparation of this, the first published results of the collaboration between the Ministry of Agriculture, Fisheries and Food and the Royal Society of Chemistry.

The early work on the collection and evaluation of literature data was done by Mrs L M Sivell. In addition, important contributions were made by Mr R W Wenlock, Dr H A Tyler and Miss H F Crawley, all of whom also helped in the design of the sampling schemes for the chemical analysis of the cereal products. The analyses were performed at the Laboratory of the Government Chemist under the able direction of Dr J R Cooke and subsequently Mr M V Meech and Mr I Lumley. This joint project between the Ministry and the Royal Society of Chemistry is in large measure the result of the encouragement and enthusiasm of Mr D D Singer.

We are indebted to Dr H N Englyst for advice and unpublished data on dietary fibre components, and to the Kellogg Company of Great Britain Ltd, Weetabix Ltd, General Foods and other cereal product manufacturers for information on their products. The Good Housekeeping Institute and the Kellogg Company kindly provided the cover photographs.

The final organisation of this book was overseen by a committee which, besides the authors, included Dr C S Berry (Flour Milling and Baking Research Association, Chorleywood), Miss P Brereton (Northwick Park Hospital, Harrow), Dr A M Fehily (MRC Epidemiology Unit, Cardiff), Miss A A Paul (MRC Dunn Nutritional Laboratory, Cambridge) and Dr D A T Southgate (AFRC Institute of Food Research, Norwich).

Last but not least, we would like to express our gratitude to the many people in the Ministry of Agriculture Fisheries and Food, the Royal Society of Chemistry, the Laboratory of the Government Chemist and elsewhere, who were involved in the production of this book.

INTRODUCTION

The fourth edition of McCance and Widdowson's *The Composition of Foods* (Paul and Southgate, 1978) is an essential text for those who need to know the nutritional value of foods consumed in Britain. In the last ten years, however, many new fresh and processed foods have become widely available, and the composition of many existing foods has changed. In addition, new analytical methods have resulted in more detailed figures on food composition. There is therefore a growing need for these tables to be updated.

The production of a completely new edition of the tables would have taken so long that it has been agreed that separate sections should instead be revised and published as supplements to the fourth edition as they were completed. The first supplement gave more details of the amino acid and fatty acid composition of foods (Paul *et al.*, 1980) and the second dealt with the foods commonly eaten by the immigrant population in the UK (Tan *et al.*, 1985). This third supplement differs from these in that it is intended to replace completely the Cereals section from the fourth edition.

Methods
A large number of new cereal products were considered for inclusion, and this supplement now contains 360 foods instead of the original 121. The selection of nutrient values has followed the general principles used in the preparation of the fourth edition - that is, using a combination of direct analysis and appropriate values from the published literature, together with some carefully selected values from manufacturers. Many of these have been combined in the recipes, which make up more than one-third of the total number of foods.

Literature values

Nutrient values from the literature were taken from studies that included full details of the samples and their preparation; where suitable methods of analysis had been used; and where the results were presented in sufficient detail. Most were from the UK because the composition of cereal products and even of bread-making and other flours can vary considerably between countries. Furthermore, flours in the UK are required by law (The Bread and Flour Regulations, 1984) to contain at least 1.65 mg iron, 0.24 mg thiamin and 1.60 mg niacin per 100 g, so these nutrients are added to all white flours and most brown flours in this country. Calcium carbonate must also be added to all flours except wholemeal and certain self-raising flours at a rate equivalent to 94–156 mg calcium per 100 g flour.

By analysis

Where a review of the literature (and, in some cases, of manufacturers' information) showed that little or nothing was known of the nutrients in an important food, arrangements were made for its direct analysis by the Laboratory of the Government Chemist. Detailed sampling and analytical protocols were devised for each item. Most of the foods were bought from a variety of shops in the London area, but breads were sampled regionally (Wenlock *et al.*, 1983) and the flours were nationally representative (Wenlock, 1982).

The analytical methods were, in general, as described in *The Composition of Foods* (Paul and Southgate, 1978) or in the supplement on *Immigrant Foods* (Tan *et al.*, 1985). In addition, individual sugars were determined colorimetrically (Southgate *et al.*, 1978) or by high-performance liquid chromatography, and total fibre and the fibre fractions were determined by the methods of Southgate (Wenlock *et al.*, 1985) and Englyst *et al.* (1988).

Arrangement of the tables

Food groups

The foods have been arranged alphabetically within the following groups: flours, grains, and starches; rice; pasta; breads; rolls; breakfast cereals; infant foods; biscuits; cakes; pastry; buns and pastries; puddings; and savoury cereal-based dishes. This is broadly as in the fourth edition, except that some of the groups have been subdivided. A few non-cereal foods such as sago, tapioca and soya flour have also been included for completeness. Some of the assignments will inevitably appear to be rather arbitrary, so a combined index and food coding list has been provided to help in locating specific foods.

Numbering system

As in the first four editions of *The Composition of Foods* the foods have been re-numbered in sequence from No. 1 (Arrowroot) to, in this supplement, No. 360 (Yorkshire pudding made with skimmed milk). We intend to number each new supplement in a similar way, so we have added a unique two digit prefix to help distinguish between each type of food. For all cereal products this code is '11'; so for arrowroot the full code number, and the one which will be used in Nutrient Databank applications, is 11-001, and the full code number for the first food in the next supplement (which will replace the Milk and Milk Product section of the fourth edition) will be 12-001.

Description and number of samples

The information given under this heading indicates the number and nature of the samples taken for analysis. The major sources of values that have been derived from the literature or from manufacturers' information, or by calculation, are also indicated under this heading. The manufacturer's name is given where this helps to identify a product. Where the calculation is from a recipe, the ingredients and details of the cooking method are given in the Appendix to this supplement. For foods that were analysed, a number of samples were purchased from different shops, supermarkets, or other retail outlets. The samples were not analysed separately but, as for previous editions, pooled before analysis. When the composite sample was made up from different brands of a food, the numbers of the individual brands purchased was related to their relative shares of the retail market. The different brands may or may not have been purchased at the same shop.

Because of the natural variability in composition between samples of the same or similar foods, apparent differences between, for example, some of the white breads in this supplement may be as much due to sampling or analytical variations as to real differences between the foods. For this reason, manufacturers however, should analyse each of their own products for nutrition labelling purposes; if the food is sufficiently similar to one described in these tables, the value quoted herein may be used as a guide.

Nutrients

The presentation of the nutrients has been changed slightly, in particular to accommodate the greater amount of information on carbohydrates and dietary fibre fractions appropriate to a description of cereal products.

Proximates:—The first of the four pages for each food shows the proximates (water, total nitrogen, protein, fat, available carbohydrate as its monosaccharide equivalent), together with energy expressed both in kilocalories and kilojoules. Protein was derived from the nitrogen values by multiplying them by the factors in **Table 1**, and the energy values were derived by multiplying the amounts of protein, fat and carbohydrate by the factors in **Table 2**.

Table 1:— **Factors used for converting total nitrogen to protein**

Wheat bran	6.31
Wholemeal wheat products	5.83
Other wheat products	5.70
Rice	5.95
Barley, oats, rye	5.83
Maize	6.25
Soya	5.71

Table 2:– **Energy conversion factors**

	kcal/g	kJ/g
Protein	4	17
Fat	9	37
Available carbohydrate expressed as monosaccharide	3.75	16

Carbohydrates and fibre:—The second page is devoted entirely to carbohydrates and fibre. There are separate values for starches (including dextrins but excluding 'resistant starch'); total sugars (including the wheat glucofructan *levosin*, as in the fourth edition); the individual sugars glucose, fructose, sucrose, maltose and lactose; two different estimates of the total fibre content of the food (that obtained by the Southgate method (Wenlock *et al.*, 1985) and total non-starch polysaccharides (Englyst *et al.*, 1988)); and five of the components or subdivisions of fibre which may have physiological activity. The carbohydrates are given as their monosaccharide equivalents, but the fibre columns are as the weight of the actual component. The relationships between the various forms and fractions of dietary fibre are shown in **Table 3**.

Minerals and vitamins:—The final two pages show the minerals in the same order as in the fourth edition, but with the addition of a number of values for manganese, selenium and iodine; and lastly the vitamins. The fat-soluble vitamins A, D and E (the latter as α–tocopherol equivalents, using the conversion factors given by McLaughlin and Weihrauch (1979)) are given first, followed by the eight B-vitamins, with total folate but not free folic acid, and then vitamin C. Retinol and carotenes have been combined, with 6 μg β-carotene taken as 1 μg retinol equivalent. Total pre-formed niacin and the potential nicotinic acid from the amino acid tryptophan (Trypt/60) are, however, given separately because of their importance in cereal products.

Table 3:— **Relationships between the dietary fibre fractions**

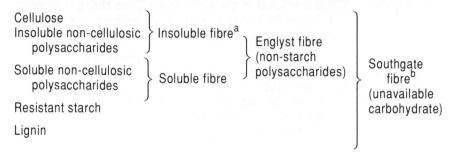

[a]Some methods of analysis also include lignin.
[b]Fibre determined by the Southgate method may differ from the sum of the fractions shown, because it can include starch which is not necessarily the same as resistant starch measured by the Englyst method.

Finally, there are tables showing the phytic acid and cholesterol content of a more limited range of cereal products.

All values are given per 100 grams of the edible portion of the food as described. The source of any individual value can be provided on request.

Recipes

The recipes in this supplement were based on a variety of sources and differ from those used in the fourth edition. The nutritional values assigned to the cereal components were from this supplement, and, where possible, updated information was used for the other ingredients too. Except where there was a (measured) uptake of fat, losses (or occasionally gains) of weight were assumed to consist only of water, and for most recipes were determined experimentally. Losses of labile vitamins were estimated from the table in the fourth edition which is reproduced below. The percentage loss of these vitamins from other ingredients was assumed to be the same as from the basic cereal ingredients, and losses of vitamin C (for example in fruit pies) were assigned individually and ranged from 10% to 70%.

Table 4:— **Percentage losses of vitamins in cereals during cooking**

	Boiling	Baking	Toasting
Thiamin	40	25*	15
Riboflavin	40	15	
Niacin	40	5	
Vitamin B_6	40	25	
Folate	50	50	
Pantothenate	40	25	
Biotin	40	0	

*15% in breadmaking

A more comprehensive description of the factors to be taken into account both in the preparation and the proper use of food composition tables is given by Paul and Southgate in their introduction to *The Composition of Foods*. Users of the present supplement would be well advised to take them to heart.

References to Introductory text

Englyst, H. N., Bingham, S., Collinson, E., Runswick, S., and Cummings, J. H. (1988) Dietary fibre content of cereal products. *J. Hum. Nutr. Diet.* (in press)

McClaughlin, P. J. and Weihrauch, J. L. (1979) Vitamin E content of foods . *J. Am. Diet. Assoc.* **75,** 647-665

Paul, A. A. and Southgate, D. A. T. (1978) *McCance and Widdowson's The Composition of Foods Fourth edition*, HMSO, London

Paul, A. A., Southgate, D. A. T. and Russell, J. (1980) *First supplement to McCance and Widdowson's The Composition of Foods: Amino acid composition (mg per 100g food) and fatty acid composition (g per 100g food),* HMSO, London

Southgate, D. A. T., Paul, A. A., Dean, A. C., and Christie, A. A. (1978) Free sugars in foods. *J. Hum. Nutr.* **32**, 335-347

Tan, S. P., Wenlock, R. W. and Buss, D. H. (1985) *Second supplement to McCance and Widdowson's The Composition of Foods: Immigrant foods.* HMSO, London

The Bread and Flour Regulations (1984) Statutory Instrument No. 1304. HMSO, London

Wenlock, R. W. (1982) The nutrient content of UK wheat flours between 1957 and 1980. *J. Sci. Food Agric.* **33**, 1310-1318

Wenlock, R. W., Sivell, L. M. and Agater, I. B. (1985) Dietary fibre fractions in cereal and cereal-containing products in Britain *J. Sci. Food Agric.* **36**, 113-121

Wenlock, R. W., Sivell, L. M., King, R. T., Scuffam, D. and Wiggins, R. A. (1983) The nutritional composition of British bread - a nationwide study. *J. Sci. Food Agric.* **34**, 1302-1318

Symbols and abbreviations used in the tables

Symbols

0	None of the nutrient is present
Tr	Trace
N	The nutrient is present in significant quantities but there is no reliable information on the amount
( )	Estimated value
Italic text	Starch, carbohydrate or energy values taken from sources other than new analytical data or the fourth edition of *The Composition of Foods*. These starch and carbohydrate values have been obtained by difference.

Abbreviations

Gluc	Glucose
Fruct	Fructose
Sucr	Sucrose
Malt	Maltose
Lact	Lactose
Trypt	Tryptophan

Flours, grains and starches

Composition of food per 100g

No. 11-	Food	Description and main data sources	Water g	Total nitrogen g	Protein g	Fat g	Carbohydrate g	Energy value kcal	Energy value kJ
1	**Arrowroot**	2 samples from different shops	12.2	0.07	0.4	0.1	94.0	355	1515
2	**Barley,** pearl *raw*	2 samples from different shops; literature sources	10.6	1.35	7.9	1.7	83.6	360	1535
3	pearl *boiled*	2 samples from different shops, boiled in water	69.6	0.46	2.7	0.6	27.6	120	510
4	whole grain *raw*	Literature sources	11.7	1.82	10.6	2.1	64.0	301	1282
5	**Bran,** wheat	Analytical and literature sources	8.3	2.24	14.1	5.5	26.8	206	872
6	**Buckwheat**	Literature sources, as groats	13.2	1.30	8.1	1.5	84.9	364	1522
7	**Bulgur wheat**	Refs. 2, 3, 5	9.3	1.70	9.7	1.7	76.3	353	1478
8	**Chapati flour,** brown	1 sample, single supplier	12.2	2.02	11.5	1.2	73.7	333	1419
9	white	2 samples, different suppliers	12.0	1.72	9.8	0.5	77.6	335	1426
10	**Cornflour**	3 samples from different shops	12.5	0.09	0.6	0.7	92.0	354	1508
11	**Cornmeal** *sifted*	Analytical and literature sources, ref. 8	12.2	1.50	9.4	3.3	73.1	368	1540
12	*unsifted*	Ref. 8 (Na, K ref. 9)	12.2	1.50	9.3	3.8	71.5	353	1477
13	**Custard powder**	Taken as cornflour except Na, Cl and Cu	12.5	0.09	0.6	0.7	92.0	354	1508
14	**Gari**	Cassava, fermented, dried. Ref. 8	12.6	0.16	1.0	1.1	82.7	351	1469
15	**Hominy/maize grits** *raw*	Ref. 5	12.0	1.40	8.7	0.8	77.7	362	1515
16	**Millet flour**	Ref. 9 (foxtail millet)	13.3	1.00	5.8	1.7	75.4	354	1481
17	**Oatmeal** *raw*	Analytical and literature sources	8.9	2.12	12.4	8.7	72.8	401	1698
18	quick cook *raw*	10 samples, 8 brands	8.2	1.92	11.2	9.2	66.0	375	1587

Flours, grains and starches

Carbohydrate fractions, g per 100g

No. 11-	Food	Starch	Total sugars	Individual sugars					Dietary fibre		Fibre fractions				
				Gluc	Fruct	Sucr	Malt	Lact	Southgate method	Englyst method	Cellulose	Non-cellulosic polysaccharide Soluble	Insoluble	Lignin	Resistant starch
1	**Arrowroot**	94.0	Tr	Tr	(0)	Tr	(0)	0	N	0.1	Tr	Tr	Tr	0	0
2	**Barley,** pearl raw	83.6	Tr	Tr	Tr	Tr	Tr	0	5.9	N	N	N	N	N	N
3	pearl boiled	27.6	Tr	Tr	Tr	Tr	Tr	0	2.0	N	N	N	N	N	N
4	whole grain raw	62.2	1.8	0.1	0.1	1.0	0.5	0	N	14.8	3.5	4.0	7.3	N	N
5	**Bran,** wheat	23.0	3.8	0.2	0.1	3.5	0.1	0	39.6	36.4	7.2	3.3	26.1	3.2	0
6	**Buckwheat**	84.5	0.4	Tr	Tr	Tr	Tr	0	N	2.1	0.5	1.0	0.6	N	0.3
7	**Bulgur wheat**	N	N	N	N	N	N	(0)	N	N	N	N	N	N	N
8	**Chapati flour,** brown	70.5	3.2[a]	N	N	N	N	(0)	10.3	N	N	N	N	0.5	N
9	white	75.5	2.1[a]	N	N	N	N	(0)	4.1	N	N	N	N	0.5	N
10	**Cornflour**	92.0	Tr	Tr	(0)	Tr	(0)	0	N	0.1	Tr	0.1	Tr	0	0.1
11	**Cornmeal** sifted	N	N	N	N	N	N	(0)	4.4	2.2	1.0	N	N	0.2	N
12	unsifted	N	N	N	N	N	N	(0)	N	N	N	N	N	N	N
13	**Custard powder**	92.0	Tr	Tr	(0)	Tr	(0)	0	N	(0.1)	Tr	Tr	Tr	0	(0.1)
14	**Gari**	N	N	N	N	N	N	(0)	N	N	N	N	N	N	N
15	**Hominy/maize grits** raw	N	N	N	N	N	N	(0)	N	N	N	N	N	N	N
16	**Millet flour**	N	N	N	N	N	N	(0)	N	N	N	N	N	N	N
17	**Oatmeal** raw	72.8	Tr	Tr	Tr	Tr	Tr	0	6.3	6.8	0.5	4.0	2.3	(1.5)	0.2
18	quick cook raw	64.9	1.1	Tr	Tr	0.8	0.3	0	6.8	7.1	0.6	4.1	2.4	1.5	0.4

[a] Includes the glucofructan levosin

9

Flours, grains and starches

No. 11-	Food	Na	K	Ca	Mg	P	Fe	Cu	Zn	S	Cl	Mn	Se	I
						mg							µg	
1	Arrowroot	5	18	7	8	27	2.0	0.22	Tr	2	7	N	N	N
2	Barley, pearl *raw*	3	270	20	65	210	3.0	0.40	2.1	120	110	1.30	(1)	N
3	pearl *boiled*	1	92	7	22	71	1.0	0.14	0.7	37	41	0.44	Tr	N
4	whole grain *raw*	4	560	50	91	380	6.0	0.30	3.3	N	23	1.70	(1)	7
5	Bran, wheat	28	1160	110	520	1200	12.9	1.34	16.2	65	150	9.00	(2)	N
6	Buckwheat	1	220	12	48	150	2.0	0.70	2.6	N	N	1.60	9	N
7	Bulgur wheat	5	290	44	140	330	4.9	0.56	N	140	22	N	N	N
8	Chapati flour, brown	39	280	86	69	250	3.4	0.33	2.1	N	67	2.00	N	N
9	white	15	200	84	29	140	2.5	0.25	1.3	N	68	0.98	N	N
10	Cornflour	52	61	15	7	39	1.4	0.13	0.3	1	71	N	N	N
11	Cornmeal *sifted*	Tr	170	3	44	120	1.3	0.15	1.0	N	N	0.20	N	N
12	*unsifted*	(1)	(280)	17	N	220	4.2	N	N	N	N	N	N	N
13	Custard powder	320	61	15	7	39	1.4	0.05	0.3	1	480	N	N	N
14	Gari	N	N	45	N	79	1.6	N	N	N	N	N	N	N
15	Hominy/maize grits *raw*	1	80	4	7	73	N	N	N	N	N	N	N	N
16	Millet flour	21	370	40	20	260	N	N	N	N	N	N	N	N
17	Oatmeal *raw*	33	370	55	110	380	4.1	0.23	(3.3)	160	73	3.70	(3)	N
18	quick cook *raw*	9	350	52	110	380	3.8	0.49	3.3	210	25	3.90	3	N

Flours, grains and starches

No. 11-	Food	Retinol equiv µg	Vitamin D µg	Vitamin E mg	Thiamin mg	Ribo-flavin mg	Niacin mg	Trypt 60 mg	Vitamin B6 mg	Vitamin B12 µg	Folate µg	Panto-thenate µg	Biotin µg	Vitamin C mg
1	**Arrowroot**	0	0	Tr	Tr	Tr	Tr	0.1	Tr	0	Tr	Tr	Tr	0
2	**Barley**, pearl *raw*	0	0	0.40	0.12	0.05	2.5	2.3	0.22	0	20	0.5	N	0
3	pearl *boiled*	0	0	0.10	0.02	0.01	0.5	0.8	0.04	0	3	0.1	Tr	0
4	whole grain *raw*	0	0	0.90	0.31	0.10	5.2	2.6	0.56	0	50	0.7	10	0
5	**Bran**, wheat	0	0	2.60	0.89	0.36	29.6	3.0	1.38	0	260	2.4	45	0
6	**Buckwheat**	N	0	Tr	0.28	0.07	2.8	1.7	0.40	0	N	1.2	N	0
7	**Bulgur wheat**	0	0	N	0.48	0.11	4.5	1.6	N	0	N	N	N	0
8	**Chapati flour**, brown	0	0	N	0.26	0.05	3.8	2.4	0.29	0	29	N	N	0
9	white	0	0	N	0.36	0.06	1.9	2.0	0.17	0	20	N	N	0
10	**Cornflour**	0	0	Tr	Tr	Tr	Tr	0.1	Tr	0	Tr	Tr	Tr	0
11	**Cornmeal** *sifted*	(4)	0	Tr	0.26	0.08	1.0	1.0	N	0	N	N	N	(3)
12	*unsifted*	4	0	Tr	0.30	0.08	1.8	1.0	N	0	N	N	N	3
13	**Custard powder**	0	0	Tr	Tr	Tr	Tr	0.1	Tr	0	Tr	Tr	Tr	0
14	**Gari**	N	0	N	0.08	0.03	1.0	0.2	N	0	N	N	N	0
15	**Hominy/maize grits** *raw*	44	0	N	0.13	0.04	1.2	0.9	N	0	N	N	N	0
16	**Millet flour**	0	0	Tr	0.68	0.19	1.6	1.2	N	0	N	N	N	0
17	**Oatmeal** *raw*	0	0	1.70	0.50	0.10	1.0	2.8	0.12	0	(60)	1.0	20	0
18	quick cook *raw*	0	0	1.50	0.90	0.09	0.8	2.6	0.33	0	60	1.2	21	0

Flours, grains and starches *continued*

Composition of food per 100g

No. 11-	Food	Description and main data sources	Water g	Total nitrogen g	Protein g	Fat g	Carbohydrate g	Energy value kcal	kJ
19	**Popcorn**, candied	Recipe	2.6	0.33	2.1	20.0	77.6	480	2018
20	plain	Recipe	0.9	0.99	6.2	42.8	48.6	592	2467
21	**Rice flour**	Ref. 9	11.8	1.08	6.4	0.8	80.1	366	1531
22	**Rye flour** *whole*	Analytical and literature sources	15.0	1.40	8.2	2.0	75.9	335	1428
23	**Sago** *raw*	2 samples from different shops	12.6	0.04	0.2	0.2	94.0	355	1515
24	**Semolina** *raw*	2 samples from different shops, coarse and fine	14.0	1.87	10.7	1.8	77.5	350	1489
25	**Soya flour** *full fat*	Analytical and literature sources	7.0	6.45	36.8	23.5	23.5	447	1871
26	*low fat*	Analytical and literature sources	7.0	7.94	45.3	7.2	28.2	352	1488
27	**Tapioca** *raw*	4 varieties, medium pearl, seed pearl, coarse and flake	12.2	0.07	0.4	0.1	95.0	359	1531
28	**Wheat flour**, brown	VFSS, 1977-81, and literature sources	14.0	2.20	12.6	1.8	68.5	323	1377
29	patent	Mixed sample	14.1	1.89	10.8	1.3	78.0	347	1480
30	white *breadmaking*	Data from Voluntary Flour Sampling Scheme (VFSS), 1977-81 plus literature sources.	14.0	2.02	11.5	1.4	75.3	341	1451
31	white *plain*		14.0	1.64	9.4	1.3	77.7	341	1450
32	white *self-raising*	Biscuit and cake flours are similar in composition to plain flour	14.0	1.56	8.9	1.2	75.6	330	1407
33	wholemeal		14.0	2.18	12.7	2.2	63.9	310	1318
34	**Wheatgerm**	Literature sources	11.7	4.54	26.7	9.2	(44.7)	302	1276

Flours, grains and starches *continued*

Carbohydrate fractions, g per 100g

No. 11-	Food	Starch	Total sugars	Gluc	Fruct	Sucr	Malt	Lact	Southgate method	Englyst method	Cellulose	Soluble	Insoluble	Lignin	Resistant starch
									Dietary fibre		**Fibre fractions**	**Non-cellulosic polysaccharide**			
19	**Popcorn**, candied	15.5	62.1	Tr	Tr	62.1	0	0	N	N	N	N	N	N	N
20	plain	47.6	1.0	0.1	0.*	0.9	0	0	N	N	N	N	N	N	N
21	**Rice flour**	80.1	N	N	N	N	N	(0)	N	2.0	0.5	N	N	N	Tr
22	**Rye flour** *whole*	75.9	N	N	N	N	N	0	N	11.7	1.2	3.9	6.6	1.9	0.2
23	**Sago** *raw*	94.0	Tr	Tr	(0)	(0)	0	0	N	0.5	0.3	0.2	0	N	0.6
24	**Semolina** *raw*	77.5	Tr	Tr	Tr	Tr	(0)	0	(3.6)	2.1	0.3	1.0	0.9	N	0.4
25	**Soya flour** *full fat*	12.3	11.2	N	N	N	N	0	10.7	11.2	2.2	5.2	3.8	N	N
26	*low fat*	14.8	13.4	N	N	N	N	N	13.3	(13.5)	(2.7)	(6.3)	(4.6)	N	N
27	**Tapioca** *raw*	95.0	Tr	Tr	(0)	Tr	(0)	0	N	0.4	0.1	0.1	0.2	N	N
28	**Wheat flour**, brown	66.8	1.7[a]	Tr	Tr	0.7	0	0	7.0	6.4	0.8	2.1	3.4	0.2	0.3
29	patent	76.6	1.4[a]	N	N	N	N	(0)	N	N	N	N	N	Tr	N
30	white *breadmaking*	73.9	1.4[a]	Tr	Tr	0.3	Tr	0	3.7	(3.1)	(0.1)	(1.5)	(1.5)	Tr	(0.3)
31	white *plain*	76.2	1.5[a]	Tr	Tr	0.3	0.2	0	3.6	3.1	0.1	1.5	1.5	Tr	0.3
32	white *self-raising*	74.3	1.3[a]	Tr	Tr	0.2	0	0	4.1	(3.1)	(0.1)	(1.5)	(1.5)	Tr	(0.3)
33	wholemeal	61.8	2.1[a]	0.1	Tr	1.0	0	0	8.6	9.0	1.4	2.0	5.6	0.3	0.3
34	**Wheatgerm**	(28.7)	(16.0)	(0.7)	(0.5)	(14.8)	0	0	N	15.6	2.7	3.2	9.7	N	0.2

[a] Includes the glucofructan levosin

Flours, grains and starches *continued*

Inorganic constituents per 100g

No. 11-	Food	Na	K	Ca	Mg	P	mg Fe	Cu	Zn	S	Cl	Mn	µg Se	µg I
19	**Popcorn**, candied	56	75	6	26	58	0.4	N	0.7	N	101	0.10	N	3
20	plain	4	220	10	81	170	1.1	N	1.7	N	8	0.32	N	2
21	**Rice flour**	5	240	24	23	130	1.9	0.20	N	N	Tr	N	N	N
22	**Rye flour** *whole*	(1)	410	32	92	360	2.7	0.42	3.0	N	N	0.68	N	N
23	**Sago** *raw*	3	5	10	3	29	1.2	0.03	N	1	13	N	N	N
24	**Semolina** *raw*	12	170	18	32	110	1.0	0.15	(0.6)	92	71	(0.60)	N	N
25	**Soya flour** *full fat*	9	1660	210	240	600	6.9	2.92	3.9	N	110	2.30	9	N
26	*low fat*	14	2030	240	290	640	9.1	3.12	3.2	N	N	2.90	(11)	N
27	**Tapioca** *raw*	4	20	8	2	30	0.3	0.07	N	4	13	N	N	N
28	**Wheat flour**, brown	4	250	130[a]	80	230	3.2[a]	0.32	1.9	N	45	1.90	N	N
29	patent	3	100	110	19	89	1.7	0.11	N	110	60	N	N	N
30	white breadmaking	3	130	140[b]	31	120	2.1[b]	0.18	0.9	110	62	0.68	42	N
31	white *plain*	3	150	140[b]	20	110	2.0[b]	0.15	0.6	N	81	0.60	4	10
32	white *self-raising*	360[c]	150	350[c]	20	450[c]	2.0[b]	0.17	0.6	N	88	0.62	4	10
33	wholemeal	3	340	38	120	320	3.9	0.45	2.9	N	38	3.14	53	N
34	**Wheatgerm**	5	950	55	270	1050	8.5	0.90	17.0	250	80	12.30	(3)	N

[a] These are values for fortified flour. Unfortified brown flour would contain about 20mg Ca and 2.5mg Fe per 100g
[b] These are values for fortified flour. Unfortified white flours would contain about 15mg Ca and 1.5mg Fe per 100g
[c] The amount present will depend on the nature and level of the raising agent used

14

Flours, grains and starches *continued*

No. 11-	Food	Retinol equiv µg	Vitamin D µg	Vitamin E mg	Thiamin mg	Ribo-flavin mg	Niacin mg	Trypt 60 mg	Vitamin B6 mg	Vitamin B12 µg	Folate µg	Panto-thenate µg	Biotin µg	Vitamin C mg
19	**Popcorn**, candied	200	0.05	3.75	0.06	0.04	0.3	0.2	0.07	0	3	0.1	1	0
20	plain	410	0	11.03	0.18	0.11	1.0	0.7	0.20	0	9	0.3	4	0
21	**Rice flour**	0	0	N	0.10	0.05	2.1	1.4	0.20	0	N	N	N	0
22	**Rye flour** *whole*	0	0	1.60	0.40	0.22	1.0	1.6	0.35	0	78	1.0	6	0
23	**Sago** *raw*	0	0	Tr	Tr	Tr	Tr	Tr	Tr	0	Tr	Tr	Tr	0
24	**Semolina** *raw*	0	0	Tr	(0.10)	(0.03)	(0.7)	2.2	(0.15)	0	(22)	(0.3)	(1)	0
25	**Soya flour** *full fat*	N	0	1.50	0.75	0.28	2.0	8.6	0.46	0	345	1.6	N	0
26	*low fat*	N	0	N	0.90	0.29	2.4	10.6	0.52	0	410	1.8	N	0
27	**Tapioca** *raw*	0	0	Tr	Tr	Tr	Tr	0.1	Tr	0	Tr	Tr	Tr	0
28	**Wheat flour**, brown	0	0	0.60	0.39[a]	0.07	4.0[a]	2.6	(0.30)	0	51	(0.4)	(3)	0
29	patent	0	0	Tr	(0.32)[b]	(0.02)[b]	(2.0)[b]	2.2	(0.10)	0	(10)	(0.3)	(1)	0
30	white *breadmaking*	0	0	(0.30)	0.32[b]	0.03	2.0[b]	2.3	0.15	0	31	0.3	1	0
31	white *plain*	0	0	0.30	0.31[b]	0.03	1.7[b]	1.9	0.15	0	22	0.3	1	0
32	white *self-raising*	0	0	(0.30)	0.30[b]	0.03	1.5[b]	1.8	0.15	0	19	0.3	1	0
33	wholemeal	0	0	1.40	0.47	0.09	5.7	2.5	0.50	0	57	0.8	7	0
34	**Wheatgerm**	0	0	22.00	2.01	0.72	4.5	5.3	3.30	0	331	1.9	25	0

[a]These are levels for fortified flour. Unfortified brown flour would contain 0.30mg thiamin and 1.7mg niacin per 100g
[b]These are levels for fortified flour. Unfortified white flours would contain 0.10mg thiamin and 0.7mg niacin per 100g

Rice

Composition of food per 100g

No. 11-	Food	Description and main data sources	Water g	Total nitrogen g	Protein g	Fat g	Carbohydrate g	Energy value kcal	kJ
35	**Brown rice** *raw*	5 assorted samples	13.9	1.10	6.7	2.8	81.3	357	1518
36	*boiled*	Water content weighed, other nutrients calculated from raw	66.0	0.43	2.6	1.1	32.1	141	597
37	**Red rice** *raw*	Ref. 9	13.2	1.24	7.4	1.6	76.0	354	1481
38	*boiled*	Calculation from raw	78.8	0.30	1.8	0.4	18.5	80	341
39	**Savoury rice** *raw*	10 samples, 5 varieties, meat and vegetable	7.0	1.41	8.4	10.3	77.4	415	1755
40	*cooked*	Calculation from raw, boiled in water	68.7	0.48	2.9	3.5[a]	26.3	142	599
41	**White rice**, basmati *raw*	Ref. 1	10.5	1.30	7.4	0.5	79.8	359	1502
42	easy cook *raw*	10 samples, 9 different brands, parboiled	11.4	1.23	7.3	3.6	85.8	383	1630
43	easy cook *boiled*	Calculation from raw	68.0	0.44	2.6	1.3	30.9	138	587
44	flakes *raw*	Ref. 4	12.6	1.16	6.6	1.2	77.5	346	1448
45	fried	Recipe	70.3	0.37	2.2	3.2	25.0	131	554
46	glutinous *raw*	Ref. 9	13.9	1.41	8.4	1.6	74.9	359	1502
47	glutinous *boiled*	Calculation from raw	83.0	0.30	1.7	0.3	14.7	65	275
48	parboiled	Refs. 5, 9	12.4	1.13	6.7	1.0	79.3	364	1523
49	polished *raw*	5 samples from different shops; literature sources	11.7	1.09	6.5	1.0	86.8	361	1536
50	polished *boiled*	5 samples boiled in water; literature sources	69.9	0.37	2.2	0.3	29.6	123	522

[a] Calculated assuming water only was added; savoury rice cooked with added fat contains approximately 8.8g fat per 100g

16

Carbohydrate fractions, g per 100g

No. 11-	Food	Starch	Total sugars	Individual sugars					Dietary fibre		Fibre fractions				
				Gluc	Fruct	Sucr	Malt	Lact	Southgate method	Englyst method	Cellulose	Non-cellulosic polysaccharide Soluble	Insoluble	Lignin	Resistant starch
35	**Brown rice** *raw*	80.0	1.3	0.5	Tr	0.8	0	0	3.8	1.9	0.6	Tr	1.3	0.1	N
36	*boiled*	31.6	0.5	0.2	Tr	0.3	0	0	1.5	0.8	0.2	Tr	0.5	Tr	N
37	**Red rice** *raw*	N	N	N	Tr	N	(0)	(0)	N	N	N	N	N	N	N
38	*boiled*	N	N	N	Tr	N	(0)	(0)	N	N	N	N	N	N	N
39	**Savoury rice** *raw*	73.8	3.6	0.2	0.5	2.5	0.2	0.1	4.0	N	N	N	N	0.3	N
40	*cooked*	25.1	1.2	0.1	0.2	0.9	0.1	Tr	1.3	1.4	0.4	N	N	0.1	0.3
41	**White rice**, basmati *raw*	79.8	Tr	Tr	Tr	Tr	(0)	(0)	N	N	N	N	N	N	N
42	easy cook *raw*	85.8	Tr	Tr	Tr	Tr	0	0	2.7	0.4	0.2	Tr	0.2	0.6	N
43	easy cook *boiled*	30.9	Tr	Tr	Tr	Tr	0	0	1.0	0.1	Tr	Tr	0.1	0.2	N
44	flakes *raw*	N	N	N	Tr	N	(0)	(0)	N	N	N	N	N	N	N
45	*fried*	23.1	1.9	0.9	0.5	0.4	0	0	1.2	0.6	0.2	0.3	0.1	N	N
46	glutinous *raw*	N	N	N	Tr	N	(0)	(0)	N	N	N	N	N	N	N
47	glutinous *boiled*	N	N	N	Tr	N	(0)	(0)	N	N	N	N	N	N	N
48	parboiled	79.3	Tr	Tr	Tr	Tr	0	0	(2.2)	(0.5)	N	N	N	N	N
49	polished *raw*	86.8	Tr	Tr	Tr	Tr	0	0	2.2	0.5	0.2	Tr	0.3	(0.6)	Tr
50	polished *boiled*	29.6	Tr	Tr	Tr	Tr	0	0	0.8	0.2	0.1	Tr	0.1	(0.2)	Tr

No. 11-	Food	Na	K	Ca	Mg	P	Fe	Cu	Zn	S	Cl	Mn	Se	I
						mg							µg	
35	**Brown rice** raw	3	250	10	110	310	1.4	0.85	1.8	90	230	2.30	(2)	N
36	boiled	1	99	4	43	120	0.5	0.33	0.7	36	91	0.90	Tr	N
37	**Red rice** raw	2	190	18	N	190	1.2	N	N	N	N	N	N	N
38	boiled	Tr	48	4	N	47	0.3	N	N	N	N	N	N	N
39	**Savoury rice** raw	1440	340	73	45	200	1.5	0.14	1.3	160	2520	1.24	N	N
40	cooked	490	110	25	15	67	0.5	0.05	0.4	55	860	0.42	N	N
41	**White rice**, basmati raw	N	N	19	N	73	1.3	N	N	N	N	N	N	N
42	easy cook raw	4	150	51	32	150	0.5	0.37	1.8	89	10	1.20	10	(14)
43	easy cook boiled	1	54	18	11	54	0.2	0.13	0.7	32	4	0.20	4	5
44	flakes raw	N	N	20	N	N	8.0	N	N	N	N	N	N	N
45	fried	56	85	7	5	38	0.3	0.06	0.5	33	98	0.28	N	N
46	glutinous raw	3	280	16	17	130	1.2	0.28	2.2	N	N	1.10	N	N
47	glutinous boiled	1	55	3	3	26	0.2	0.05	0.4	N	N	0.22	N	N
48	parboiled	2	150	7	90	130	1.2	0.34	2.0	100	N	3.90	(14)	N
49	polished raw	6	110	4	13	100	0.5	0.18	1.3	78	27	0.87	(10)	(14)
50	polished boiled	2	38	1	4	34	0.2	0.06	0.5	27	9	0.30	(4)	5

No. 11-	Food	Retinol equiv µg	Vitamin D µg	Vitamin E mg	Thiamin mg	Ribo-flavin mg	Niacin mg	Trypt 60 mg	Vitamin B6 mg	Vitamin B12 µg	Folate µg	Panto-thenate µg	Biotin µg	Vitamin C mg
35	**Brown rice** raw	0	0	0.80	0.59	0.07	5.3	1.5	N	0	49	N	N	0
36	boiled	0	0	0.30	0.14	0.02	1.3	0.6	N	0	10	N	N	0
37	**Red rice** raw	0	0	N	0.30	0.10	4.2	1.7	N	0	N	N	N	0
38	boiled	0	0	N	0.04	0.01	0.6	0.4	N	0	N	N	N	0
39	**Savoury rice** raw	N	N	N	0.46	0.06	5.2	1.9	0.37	Tr	25	N	N	0
40	cooked	N	N	N	0.10	0.01	1.1	0.6	0.07	Tr	4	N	N	0
41	**White rice**, basmati raw	0	0	N	N	N	N	N	N	0	N	N	N	0
42	easy cook raw	0	0	(0.10)	0.41	0.02	4.2	1.6	0.31	0	20	(0.6)	(3)	0
43	easy cook boiled	0	0	Tr	0.01	Tr	0.9	0.6	0.07	0	4	(0.1)	(1)	0
44	flakes raw	0	0	N	0.21	0.05	4.0	1.5	N	0	N	N	N	0
45	fried	0	0	N	0.03	0.01	0.3	0.5	0.06	0	3	0.1	1	1
46	glutinous raw	0	0	N	0.16	0.06	2.4	1.9	N	0	N	N	N	0
47	glutinous boiled	0	0	N	0.02	Tr	0.3	0.4	N	0	N	N	N	0
48	parboiled	0	0	N	0.20	0.08	2.6	1.5	N	0	11	N	N	0
49	polished raw	0	0	0.10	0.08	(0.02)	1.5	1.5	0.30	0	(20)	0.6	3	0
50	polished boiled	0	0	Tr	0.01	(0.01)	0.3	0.5	0.05	0	(3)	0.2	1	0

No. 11-	Food	Description and main data sources	Water g	Total nitrogen g	Protein g	Fat g	Carbohydrate g	Energy value kcal	kJ
51	**Lasagna** *raw*	10 samples, 7 brands lasagna and cannelloni	9.7	2.09	11.9	2.0	74.8	346	1473
52	*boiled*	Calculation from raw	75.7	0.52	3.0	0.6	22.0	100	424
53	**Macaroni** *raw*	10 samples, 7 brands; literature sources	9.7	2.11	12.0	1.8	75.8	348	1483
54	*boiled*	10 samples, 7 brands boiled in water	78.1	0.52	3.0	0.5	18.5	86	365
55	**Noodles**, egg *raw*	10 samples, 8 brands	9.1	2.12	12.1	8.2	71.7	391	1656
56	egg *boiled*	10 samples, 8 brands boiled in water	84.3	0.40	2.2	0.5	13.0	62	264
57	fried	Recipe	75.1	0.34	1.9	11.5	11.3	153	638
58	plain *raw*	10 samples, 6 brands	10.6	2.05	11.7	6.2	76.1	388	1646
59	plain *boiled*	10 samples, 6 brands boiled in water	82.2	0.42	2.4	0.4	13.0	62	264
60	rice *dried*	Ref. 9	13.0	0.82	4.9	0.1	81.5	360	1506
61	**Spaghetti**, white *raw*	10 samples, 7 brands	9.8	2.11	12.0	1.8	74.1	342	1456
62	white *boiled*	10 samples, 7 brands boiled in water	73.8	0.63	3.6	0.7	22.2	104	442
63	wholemeal *raw*	10 samples, 5 brands	10.5	2.30	13.4	2.5	66.2	324	1379
64	wholemeal *boiled*	Water content weighed, other nutrients calculated from raw	69.1	0.81	4.7	0.9	23.2	113	485
65	**Vermicelli** *raw*	Ref. 2	11.7	1.40	8.7	0.4	78.3	355	1485

11-051 to 11-065

Carbohydrate fractions, g per 100g

No. 11-	Food	Starch	Total sugars	Individual sugars					Dietary fibre		Fibre fractions				
				Gluc	Fruct	Sucr	Malt	Lact	Southgate method	Englyst method	Cellulose	Non-cellulosic polysaccharide		Lignin	Resistant starch
												Soluble	Insoluble		
51	**Lasagna** raw	71.5	3.3	0.3	0.1	0.7	1.5	0	4.9	(3.1)	(0.4)	(1.6)	(0.9)	0.2	(0.3)
52	boiled	21.5	0.5	Tr	Tr	0.1	0.2	0	1.4	(0.9)	(0.1)	(0.5)	(0.3)	Tr	(0.6)
53	**Macaroni** raw	73.6	2.2	0.2	0.1	0.6	1.2	0	5.0	3.1[a]	0.4	1.6	0.9	0.1	0.3
54	boiled	18.2	0.3	Tr	Tr	0.1	0.2	0	1.5	0.9[a]	0.1	0.5	0.3	0.1	0.6
55	**Noodles,** egg raw	69.8	1.9	0.1	Tr	0.6	1.1	0	5.0	(2.9)	(0.3)	(1.5)	(1.1)	0.3	(0.3)
56	egg boiled	12.8	0.2	Tr	Tr	0.1	0.1	0	1.0	(0.6)	(0.1)	(0.3)	(0.2)	0.1	(0.5)
57	fried	10.6	0.6	0.2	0.1	0.1	0.1	0	0.9	0.5	N	N	N	0.1	N
58	plain raw	73.7	2.4	0.1	0.1	0.7	1.4	0	5.2	(2.9)	(0.3)	(1.5)	(1.1)	0.3	(0.3)
59	plain boiled	12.8	0.2	Tr	Tr	0.1	0.1	0	1.2	(0.7)	(0.1)	(0.4)	(0.3)	(0.1)	(0.5)
60	rice dried	N	N	N	N	N	N	(0)	N	N	N	N	N	N	N
61	**Spaghetti,** white raw	70.8	3.3	0.3	0.1	0.8	1.8	0	5.1	2.9	0.3	1.5	1.1	0.2	0.3
62	white boiled	21.7	0.5	Tr	Tr	0.1	0.3	0	1.8	1.2	0.1	0.6	0.5	0.2	0.5
63	wholemeal raw	62.5	3.7	0.8	0.4	1.1	1.2	0	11.5	8.4	1.6	2.0	4.8	0.6	0.3
64	wholemeal boiled	21.9	1.3	0.3	0.1	0.4	0.4	0	4.0	3.5	0.7	0.8	2.0	0.2	0.4
65	**Vermicelli** raw	N	N	N	N	N	N	(0)	N	N	N	N	N	N	N

[a]Wholemeal macaroni contains 8.3g (raw) and 2.8g (boiled) Englyst fibre per 100g

Pasta

Inorganic constituents per 100g

No. 11-	Food	Na	K	Ca	Mg	P	Fe	Cu	Zn	S	Cl	Mn	Se	I
							mg						µg	
51	**Lasagna** raw	10	230	23	48	200	1.8	0.29	1.5	170	20	0.80	(16)	Tr
52	boiled	1	23	6	12	50	0.5	0.09	0.5	43	5	0.24	(4)	Tr
53	**Macaroni** raw	11	230	25	53	180	1.6	0.30	1.5	170	20	0.86	16	Tr
54	boiled	1	25	6	14	42	0.5	0.09	0.5	40	5	0.26	4	Tr
55	**Noodles**, egg raw	180	260	28	43	200	1.5	0.24	1.3	180	180	0.79	N	N
56	egg boiled	15	23	5	8	31	0.3	0.06	0.3	32	10	0.16	N	N
57	fried	84	28	6	7	27	0.3	0.05	0.3	29	120	0.14	N	Tr
58	plain raw	2	230	23	47	160	1.5	0.42	1.5	160	20	0.79	(1)	Tr
59	plain boiled	1	13	5	8	28	0.3	0.06	0.3	32	4	0.17	Tr	Tr
60	rice dried	12	5	12	N	32	1.5	N	N	N	N	N	N	Tr
61	**Spaghetti**, white raw	3	250	25	56	190	2.1	0.32	1.5	180	25	0.94	(1)	Tr
62	white boiled	Tr	24	7	15	44	0.5	0.10	0.5	46	Tr	0.26	Tr	Tr
63	wholemeal raw	130	390	31	120	330	3.9	0.51	3.0	190	210	2.64	N	N
64	wholemeal boiled	45	140	11	42	110	1.4	0.18	1.1	67	73	0.90	N	N
65	**Vermicelli** raw	8	140	22	42	92	2.0	0.29	N	150	46	N	N	N

No. 11-	Food	Retinol equiv μg	Vitamin D μg	Vitamin E mg	Thiamin mg	Ribo-flavin mg	Niacin mg	Trypt 60 mg	Vitamin B6 mg	Vitamin B12 μg	Folate μg	Panto-thenate μg	Biotin μg	Vitamin C mg
51	**Lasagna** *raw*	Tr	Tr	Tr	0.49	0.06	2.5	2.4	0.11	Tr	37	(0.3)	(1)	0
52	*boiled*	Tr	Tr	Tr	0.05	0.01	0.4	0.6	0.01	Tr	6	Tr	Tr	0
53	**Macaroni** *raw*	0	0	Tr	0.18	0.05	2.9	2.5	0.10	(0)	23	(0.3)	(1)	0
54	*boiled*	0	0	Tr	0.03	Tr	0.5	0.6	0.01	(0)	3	Tr	Tr	0
55	**Noodles**, egg *raw*	37	0.3	N	0.26	0.10	2.2	2.5	0.10	Tr	29	N	N	0
56	egg *boiled*	2	Tr	N	0.01	0.01	0.2	0.5	0.01	Tr	1	N	N	0
57	*fried*	2	0	2.67	0.01	0.01	0.2	0.4	0.01	(0)	1	Tr	0	0
58	plain *raw*	0	0	Tr	0.37	0.04	2.4	2.4	0.13	(0)	18	(0.3)	(1)	0
59	plain *boiled*	0	0	Tr	0.02	Tr	0.3	0.5	0.01	(0)	2	Tr	Tr	0
60	rice *dried*	(0)	(0)	Tr	0.04	0.01	0.3	3.0	N	(0)	N	N	N	0
61	**Spaghetti**, white *raw*	0	0	Tr	0.22	0.03	3.1	2.5	0.17	(0)	34	(0.3)	(1)	0
62	white *boiled*	0	0	Tr	0.01	0.01	0.5	0.7	0.02	(0)	4	Tr	Tr	0
63	wholemeal *raw*	0	0	Tr	0.99	0.11	6.2	2.7	0.39	(0)	40	(0.8)	(1)	0
64	wholemeal *boiled*	0	0	Tr	0.21	0.02	1.3	1.0	0.08	(0)	7	(0.2)	Tr	0
65	**Vermicelli** *raw*	(0)	(0)	Tr	0.19	0.05	1.8	1.6	N	(0)	N	N	N	0

Breads

Composition of food per 100g

No. 11-	Food	Description and main data sources	Water g	Total nitrogen g	Protein g	Fat g	Carbohydrate g	Energy value kcal	kJ
66	**Bannocks** *made with beremeal*	Recipe, Orkney Isles	27.4	1.47	8.6	3.5	56.0	304	1275
67	*made with wheat flour*	Recipe, Orkney Isles	27.7	1.49	8.7	2.6	62.7	293	1246
68	**Breadcrumbs** *homemade*	Mean of several samples	9.7	2.03	11.6	1.9	77.5	354	1508
69	*manufactured*	10 samples, 10 different brands	6.8	1.77	10.1	2.1	78.5	354	1505
70	**Brown bread** *average*	Average of 2 types of brown bread, sliced and unsliced	39.5	1.48	8.5	2.0	44.3	218	927
71	*large, sliced*	3 types, 21 samples	41.0	1.42	8.1	1.6	44.1	212	903
72	*large, unsliced*	2 types, 13 samples	38.0	1.54	8.8	2.4	44.4	223	949
73	*toasted*	Calculated using 22% weight loss	24.4	1.82	10.4	2.1	56.5	272	1158
74	**Chapatis** *made with fat* [a]	6 samples	28.5	1.42	8.1	12.8	48.3	328	1383
75	*made without fat*	Analysed and calculated values	45.8	1.28	7.3	1.0	43.7	202	860
76	**Currant bread**	10 samples, 10 different shops	29.4	1.32	7.5	7.6	50.7	289	1220
77	*toasted*	Calculated using 12% weight loss	25.9	1.48	8.4	8.5	56.8	323	1366
78	**Granary bread**	10 samples, 10 different shops	35.4	1.62	9.3	2.7	46.3	235	999
79	**Hovis** *average*	Average of 3 types of Hovis (wheatgerm) bread	40.3	1.67	9.5	2.0	41.5	212	899
80	*large, sliced*	13 samples, different shops	41.5	1.63	9.3	1.6	41.0	205	873
81	*large, unsliced*	6 samples, different shops	40.5	1.68	9.6	1.9	41.3	210	896
82	*small, unwrapped*	8 samples, different shops	38.9	1.70	9.7	2.5	42.2	220	933
83	*toasted*	Calculated using 22% weight loss	23.5	2.04	12.1	2.6	53.2	271	1151

[a] Puris (deep fried chapatis) contain 19.1g water, 7.0g protein, 25.0g fat, and 43.3g carbohydrate

Carbohydrate fractions, g per 100g

No. 11-	Food	Starch	Total sugars	Gluc	Fruct	Sucr	Malt	Lact	Southgate method	Englyst method	Cellulose	Soluble	Insoluble	Lignin	Resistant starch
				Individual sugars					Dietary fibre		Non-cellulosic polysaccharide			Fibre fractions	
66	**Bannocks** *made with beremeal*	53.3	2.7	0.3	0.3	0.3	Tr	1.8	N	N	N	N	N	N	N
67	*made with wheat flour*	59.7	3.0	0.4	0.4	0.2	0.1	1.9	2.8	2.4	0.1	1.2	1.2	Tr	N
68	**Breadcrumbs** *homemade*	74.9	2.6	N	N	N	N	0	(5.6)	(2.2)	(0.2)	(1.3)	(0.7)	(0.6)	N
69	*manufactured*	73.5	5.0	0.2	0.5	0.3	3.9	Tr	5.3	N	N	N	N	0.2	N
70	**Brown bread** *average*	41.3	3.0	N	N	N	N	0	5.9	(3.5)	(0.5)	(1.1)	(1.8)	0.4	(0.8)
71	*large, sliced*	40.6	3.5	N	N	N	N	0	5.5	3.5	0.5	1.1	1.8	0.5	0.8
72	*large, unsliced*	42.0	2.4	N	N	N	N	0	6.2	(3.5)	(0.5)	(1.1)	(1.8)	0.3	(0.8)
73	*toasted*	52.1	4.5	N	N	N	N	0	7.1	4.5	0.6	1.4	2.3	0.6	N
74	**Chapatis** *made with fat*	46.5	1.8	N	N	N	N	0	7.0a	N	N	N	N	N	N
75	*made without fat*	42.1	1.6	N	N	N	N	0	6.4	N	N	N	N	N	N
76	**Currant bread**	36.3	14.4	6.3	6.7	0	1.2	0.2	3.8	N	N	N	N	0.3	N
77	*toasted*	40.7	16.1	7.0	7.5	0	1.4	0.2	4.2	N	N	N	N	0.3	N
78	**Granary bread**	44.1	2.2	N	N	N	N	0	6.5	4.3	0.7	2.1	1.5	0.6	0.9
79	**Hovis** *average*	39.7	1.8	N	N	N	N	0	5.1	3.3	0.4	1.3	1.6	0.4	N
80	*large, sliced*	38.9	2.1	N	N	N	N	0	5.4	(3.3)	(0.4)	(1.3)	(1.6)	0.4	N
81	*large, unsliced*	39.6	1.7	N	N	N	N	0	4.9	(3.3)	(0.4)	(1.3)	(1.6)	0.4	N
82	*small, unwrapped*	40.5	1.7	N	N	N	N	0	4.9	(3.3)	(0.4)	(1.3)	(1.6)	0.4	N
83	*toasted*	50.9	2.3	N	N	N	N	N	6.5	(4.2)	(0.5)	(1.7)	(2.0)	0.5	N

aPuris (deep fried chapatis) contain 4.4g Southgate fibre per 100g

Breads

Inorganic constituents per 100g

No. 11-	Food	Na	K	Ca	Mg	P	Fe (mg)	Cu	Zn	S	Cl	Mn	Se (µg)	I
66	**Bannocks** *made with*													
	beremeal	270	340	97	120	200	2.4	0.40	1.8	N	120	N	N	N
67	*made with wheat flour*	25	180	160	20	120	1.6	0.12	0.6	N	100	0.47	3	17
68	**Breadcrumbs** *homemade*	760	150	130	34	120	2.8	0.20	(0.7)	110	1140	(0.60)	N	N
69	*manufactured*	400	190	110	28	130	2.1	0.30	1.0	130	700	0.80	N	N
70	**Brown bread** *average*	540	170	100	53	150	2.2	0.16	1.1	85	890	1.20	N	N
71	*large, sliced*	540	160	110	48	140	2.1	0.12	1.0	(85)	910	1.03	N	N
72	*large, unsliced*	550	180	91	59	160	2.3	0.20	1.2	(85)	870	1.30	N	N
73	*toasted*	690	210	140	62	180	2.7	0.15	1.3	(110)	1170	1.30	N	N
74	**Chapatis** *made with fat*	130	160	66	41	130	2.3	0.20	1.1	N	250	(1.40)	N	29
75	*made without fat*	120	150	60	37	120	2.1	0.20	1.0	N	230	(1.24)	N	N
76	**Currant bread**	290	220	86	26	93	1.6	0.32	0.7	100	480	0.43	N	29
77	*toasted*	330	250	96	29	100	1.8	0.36	0.8	110	540	0.48	N	33
78	**Granary bread**	580	190	77	59	180	2.7	0.18	1.5	N	930	1.40	N	N
79	**Hovis** *average*	600	200	120	56	190	3.7	0.24	2.1	88	900	1.90	N	22
80	*large, sliced*	560	190	130	57	180	3.7	0.23	2.1	(88)	790	1.90	N	(22)
81	*large, unsliced*	640	200	110	56	190	3.6	0.24	2.1	(88)	1050	1.90	N	(22)
82	*small, unwrapped*	590	200	120	56	190	3.8	0.24	2.0	(88)	850	1.90	N	(22)
83	*toasted*	770	250	150	72	240	4.7	0.31	2.7	(110)	1150	2.40	N	28

No. 11-	Food	Retinol equiv µg	Vitamin D µg	Vitamin E mg	Thiamin mg	Ribo- flavin mg	Niacin mg	Trypt 60 mg	Vitamin B6 mg	Vitamin B12 µg	Folate µg	Panto- thenate µg	Biotin µg	Vitamin C mg
66	**Bannocks** *made with beremeal*	22	0.01	0.25	0.21	0.11	3.3	2.2	0.18	Tr	N	N	N	Tr
67	*made with wheat flour*	22	0.01	0.28	0.20	0.08	1.3	1.8	0.11	Tr	10	0.3	2	Tr
68	**Breadcrumbs** *homemade*	0	0	Tr	(0.20)	(0.05)	(1.8)	2.4	(0.08)	0	(25)	(0.3)	(1)	0
69	*manufactured*	N	N	N	0.54	0.08	1.8	2.1	0.07	Tr	19	N	N	0
70	**Brown bread** *average*	0	0	Tr	0.27	0.09	2.5	1.7	0.13	0	40	0.3	3	0
71	*large, sliced*	0	0	Tr	0.24	0.09	3.0	1.7	0.11	0	34	(0.3)	(3)	0
72	*large, unsliced*	0	0	Tr	0.29	0.10	3.0	1.8	0.16	0	46	(0.3)	(3)	0
73	*toasted*	0	0	Tr	0.26	0.12	3.1	2.1	0.14	0	44	(0.4)	(4)	0
74	**Chapatis** *made with fat*	N	N	N	0.26	0.04	1.7	1.7	(0.20)	0	15	(0.2)	(2)	0
75	*made without fat*	0	0	Tr	0.23	0.04	1.5	1.5	(0.18)	0	14	(0.2)	(2)	0
76	**Currant bread**	Tr	0	Tr	0.19	0.09	1.5	1.5	0.09	Tr	19	N	N	0
77	*toasted*	Tr	0	Tr	0.16	0.10	1.7	1.7	0.10	Tr	21	N	N	0
78	**Granary bread**	0	0	N	0.30	0.11	3.0	1.9	0.17	0	90	N	N	0
79	**Hovis** *average*	0	0	N	0.80	0.09	4.2	1.9	0.11	0	39	(0.3)	(2)	0
80	*large, sliced*	0	0	N	0.79	0.05	4.1	1.9	0.10	0	32	(0.3)	(2)	0
81	*large, unsliced*	0	0	N	0.81	0.13	4.3	2.0	0.11	0	43	(0.3)	(2)	0
82	*small, unwrapped*	0	0	N	0.81	0.08	4.3	2.0	0.11	0	43	(0.3)	(2)	0
83	*toasted*	0	0	N	0.87	0.12	5.4	2.6	0.14	0	50	(0.4)	(3)	0

Breads continued

Composition of food per 100g

No. 11-	Food	Description and main data sources	Water g	Total nitrogen g	Protein g	Fat g	Carbohydrate g	Energy value kcal	kJ
84	**Malt bread**	10 samples, 5 brands	25.8	1.46	8.3	2.4	56.8	268	1139
85	**Milk bread**	Recipe	33.0	1.55	9.0	8.7	48.5	296	1248
86	**Naan bread**	Recipe	28.8	1.53	8.9	12.5	50.1	336	1415
87	**Papadums** raw	10 samples, 4 brands	12.1	3.30	20.6	1.9	46.0	272	1157
88	fried	Calculated using weighed fat uptake	10.3	2.80	17.5	16.9	39.1	369	1548
89	**Paratha**	Recipe. Ref. 6	32.3	1.40	8.0	14.3	43.2	322	1356
90	**Pitta bread**, white	10 samples, 4 brands	32.7	1.61	9.2	1.2	57.9	265	1127
91	**Rye bread**	15 samples, different shops; literature sources	37.4	1.46	8.3	1.7	45.8	219	932
92	**Soda bread**	Recipe	36.0	1.32	7.7	2.5	54.6	258	1095
93	**Tortillas** made with wheat flour	Recipe	33.2	1.26	7.2	1.0	59.7	262	1114
94	**Vitbe** average	Average of 3 types of Vitbe (wheatgerm) bread	37.5	1.70	9.7	3.1	43.4	229	974
95	large, wrapped	5 samples, different shops	38.1	1.64	9.3	3.4	43.0	229	972
96	small, unwrapped	5 samples, different shops	36.5	1.75	10.0	2.9	44.1	231	983
97	small, wrapped	5 samples, different shops	37.8	1.70	9.7	3.0	43.1	227	966
98	**Wheatgerm bread** average	Average of Hovis and Vitbe breads	38.9	1.70	9.2	2.5	42.5	232	937

No. 11-	Food	Starch	Total sugars	Individual sugars					Dietary fibre		Fibre fractions				
				Gluc	Fruct	Sucr	Malt	Lact	Southgate method	Englyst method	Cellulose	Non-cellulosic polysaccharide		Lignin	Resistant starch
												Soluble	Insoluble		
84	**Malt bread**	30.7	26.1	6.1	5.8	0.3	13.5	0.4	6.5	N	N	N	N	1.5	(0.5)
85	**Milk bread**	45.8	2.7	0.3	0.3	0.2	Tr	1.8	2.6	1.9	0.1	0.9	0.9	N	N
86	**Naan bread**	44.6	5.5	0.3	0.3	2.1	Tr	2.8	2.2	1.9	0.1	0.9	0.9	N	N
87	**Papadums** raw	46.0	0	0	0	0	0	0	10.7	N	N	N	N	N	N
88	fried	39.1	0	0	0	0	0	0	9.1	N	N	N	N	N	N
89	**Paratha**	42.1	1.1	0.3	0.3	0.4	0	0	4.4	4.0	0.5	1.3	2.1	0.1	N
90	**Pitta bread**, white	55.5	2.4	N	N	N	N	0	3.9[a]	(2.2)[a]	N	N	N	N	N
91	**Rye bread**	44.0	1.8	N	N	N	N	0	5.8	4.4[b]	0.4	2.2	1.8	0.3	0.8
92	**Soda bread**	51.7	2.9	0.3	0.3	0.2	0.1	1.9	2.4	2.1	0.1	1.0	1.0	N	N
93	**Tortillas** made with wheat flour	58.6	1.1	0.4	0.4	0.2	0.1	0	2.8	2.4	0.1	1.2	1.2	Tr	N
94	**Vitbe** average	40.4	3.0	N	N	N	N	0	5.8	(3.3)	(0.4)	(1.3)	(1.6)	0.5	N
95	large, wrapped	39.9	3.1	N	N	N	N	0	6.4	(3.3)	(0.4)	(1.3)	(1.6)	0.5	N
96	small, unwrapped	41.2	2.9	N	N	N	N	0	5.4	(3.3)	(0.4)	(1.3)	(1.6)	0.4	N
97	small, wrapped	40.1	3.0	N	N	N	N	0	5.6	(3.3)	(0.4)	(1.3)	(1.6)	0.5	N
98	**Wheatgerm bread** average	40.1	2.4	N	N	N	N	0	5.4	(3.3)	(0.4)	(1.3)	(1.6)	0.4	N

[a] Wholemeal pitta bread contains 5.2g Englyst fibre and 9.0g Southgate fibre per 100g
[b] Pumpernickel contains approximately 7.5g Englyst fibre per 100g

29

Breads *continued*

Inorganic constituents per 100g

No. 11-	Food	Na	K	Ca	Mg	P	Fe	Cu	Zn	S	Cl	Mn	Se	I
							mg						µg	
84	**Malt bread**	280	280	110	45	160	2.8	0.26	1.1	64	260	0.80	N	27
85	**Milk bread**	460	170	130	30	130	1.7	0.20	0.8	81	750	0.42	26	N
86	**Naan bread**	380	180	160	28	130	1.3	0.12	0.8	N	630	0.41	25	19
87	**Papadums** *raw*	2850	880	81	210	290	13.0	0.63	2.9	N	3110	1.53	N	N
88	*fried*	2420	750	69	170	250	11.0	0.50	2.5	N	2640	1.30	N	N
89	**Paratha**	120	160	84	51	150	2.0	0.21	1.2	N	240	1.20	N	N
90	**Pitta bread**, white	520[a]	110	91[a]	24	92	1.7[a]	0.21	0.6[a]	N	830	0.44	N	N
91	**Rye bread**	580	190	80	48	160	2.5	0.18	1.3	N	1410	1.00	N	N
92	**Soda bread**	420	250	140	20	110	1.4	0.10	0.6	N	500	0.41	3	16
93	**Tortillas** *made with wheat flour*	280	110	110	17	85	1.5	0.12	0.5	N	490	0.46	3	8
94	**Vitbe** *average*	550	180	150	52	170	2.1	0.18	1.7	(88)	950	1.70	N	(22)
95	*large, wrapped*	530	180	130	51	170	2.2	0.12	1.7	(88)	850	1.70	N	(22)
96	*small, unwrapped*	550	180	110	56	170	2.0	0.26	1.7	(88)	930	1.70	N	(22)
97	*small, wrapped*	570	180	140	50	160	2.1	0.17	1.7	(88)	1060	1.60	N	(22)
98	**Wheatgerm bread** *average*	570	190	120	54	180	2.9	0.21	1.9	(88)	920	1.78	N	(22)

[a] Wholemeal pitta bread contains 460mg Na, 48mg Ca, 2.7mg Fe, and 1.8mg Zn per 100g

No. 11-	Food	Retinol equiv µg	Vitamin D µg	Vitamin E mg	Thiamin mg	Ribo- flavin mg	Niacin mg	Trypt 60 mg	Vitamin B6 mg	Vitamin B12 µg	Folate µg	Panto- thenate µg	Biotin µg	Vitamin C mg
84	**Malt bread**	Tr	(0)	N	0.45	0.13	2.8	1.7	0.11	(0)	30	N	N	0
85	**Milk bread**	84	0.62	0.84	0.22	0.13	1.8	1.8	0.11	Tr	44	0.4	5	Tr
86	**Naan bread**	97	0.21	1.38	0.19	0.10	1.2	1.8	0.08	Tr	14	0.3	2	Tr
87	**Papadums** *raw*	N	0	N	0.21	0.12	1.2	2.7	N	0	67	N	N	0
88	*fried*	N	0	N	0.13	0.09	1.0	2.3	N	0	28	N	N	0
89	**Paratha**	140	0.12	0.70	0.18	0.04	2.4	1.7	0.14	0	16	0.2	2	0
90	**Pitta bread**, white	0	0	N	0.24	0.05	1.4	1.9	N	0	21	N	N	0
91	**Rye bread**	0	0	1.20	0.29	0.05	2.3	1.7	0.09	0	24	0.5	N	0
92	**Soda bread**	22	0.01	0.24	0.17	0.08	1.1	1.6	0.09	Tr	9	0.3	1	Tr
93	**Tortillas** *made with wheat flour*	0	0	0.23	0.18	0.02	1.2	1.5	0.09	0	8	0.2	1	0
94	**Vitbe** *average*	0	0	N	0.23	0.10	1.7	2.0	0.18	0	53	(0.3)	(2)	0
95	*large wrapped*	0	0	N	0.28	0.08	1.5	1.9	0.21	0	62	(0.3)	(2)	0
96	*small, unwrapped*	0	0	N	0.19	0.10	2.1	2.0	0.16	0	33	(0.3)	(2)	0
97	*small, wrapped*	0	0	N	0.21	0.13	1.5	2.0	0.17	0	63	(0.3)	(2)	0
98	**Wheatgerm bread**, average	0	0	N	0.52	0.09	3.0	2.0	0.15	0	46	(0.3)	(2)	0

Breads *continued*

No. 11-	Food	Description and main data sources	Water g	Total nitrogen g	Protein g	Fat g	Carbohydrate g	Energy value kcal	kJ
99	**White bread** *average*	Weighted average of 5 main types of white bread	37.3	1.47	8.4	1.9	49.3	235	1002
100	*large, crusty*	27 samples, cob and bloomer	35.6	1.55	8.8	2.0	50.5	243	1031
101	*large, tin*	28 samples, split and farmhouse	37.6	1.46	8.3	1.8	48.8	232	985
102	*sliced*	42 samples, 6 batches	40.4	1.33	7.6	1.3	46.8	217	926
103	*small, unwrapped*	28 samples, 4 different types	35.4	1.54	8.8	2.2	51.2	245	1044
104	*small, wrapped*	21 samples, 3 batches	37.3	1.45	8.3	2.3	49.1	238	1012
105	*fried*	Calculated on white sliced bread using analysed fat and water changes	7.4	1.38	7.9	32.2[a]	48.5	503	2102
106	*toasted*	Calculated using water loss of 18%	27.3	1.62	9.3	1.6	57.1	265	1129
107	*French stick*	10 samples, 10 different shops	29.2	1.68	9.6	2.7	55.4	270	1149
108	Scottish batch *unwrapped*	10 samples, 10 different shops	36.5	1.52	8.7	1.3	51.1	238	1014
109	*wrapped*	10 samples, 10 different shops	38.8	1.50	8.6	1.2	48.6	227	968
110	*Vienna*	10 samples, 10 different shops	32.6	1.63	9.3	3.3	52.2	263	1115
111	*West Indian*	10 assorted samples	33.2	1.46	8.3	3.4	58.8	284	1208
112	*aerated (slimmers)*	10 samples, 10 different shops	35.3	1.62	9.2	2.2	50.9	247	1052
113	**Wholemeal bread** *average*	Average of 3 types	38.3	1.58	9.2	2.5	41.6	215	914
114	*large*	16 samples, different shops	38.6	1.54	9.0	2.6	41.9	217	920
115	*small, sliced*	15 samples, different shops	39.6	1.60	9.3	2.0	40.3	206	877
116	*small, unsliced*	12 samples, different shops	36.8	1.59	9.3	2.9	42.7	223	949
117	*toasted*	Calculated using water loss of 14.6%	27.8	1.85	10.8	2.9	48.7	252	1070

[a] The fat content depends on the conditions of frying; thin slices pick up proportionately more fat than thick ones

32

Carbohydrate fractions, g per 100g

No. 11-	Food	Starch	Total sugars	Gluc	Fruct	Sucr	Malt	Lact	Dietary fibre Southgate method	Dietary fibre Englyst method	Cellulose	Non-cellulosic polysaccharide Soluble	Non-cellulosic polysaccharide Insoluble	Lignin	Resistant starch
99	**White bread** *average*	46.7	2.6	N	N	N	N	0	3.8	(1.5)[a]	(0.1)	(0.9)	(0.5)	0.4	(0.8)
100	*large, crusty*	47.5	2.9	N	N	N	N	0	4.1	(1.5)	(0.1)	(0.9)	(0.5)	0.2	(0.8)
101	*large, tin*	46.2	2.6	N	N	N	N	0	4.0	(1.5)	(0.1)	(0.9)	(0.5)	0.3	(0.8)
102	*sliced*	43.8	3.0	N	N	N	N	0	3.7	1.5	0.1	0.9	0.5	0.5	0.8
103	*small, unwrapped*	49.0	2.2	N	N	N	N	0	3.8	(1.5)	(0.1)	(0.9)	(0.5)	0.4	(0.8)
104	*small, wrapped*	46.9	2.2	N	N	N	N	0	3.5	(1.5)	(0.1)	(0.9)	(0.5)	0.4	(0.8)
105	*fried*	45.3	3.1	N	N	N	N	0	3.8	(1.6)	(0.1)	(0.9)	(0.5)	0.5	N
106	*toasted*	53.4	3.7	N	N	N	N	0	4.5	1.8	0.1	1.1	0.6	0.6	N
107	*French stick*	53.5	1.9	N	N	N	N	0	5.1	(1.5)	(0.1)	(0.9)	(0.5)	0.6	1.0
108	*Scottish batch unwrapped*	48.5	2.6	N	N	N	N	0	3.9	(1.5)	(0.1)	(0.9)	(0.5)	0.2	(0.8)
109	*Scottish batch wrapped*	46.1	2.5	N	N	N	N	0	3.9	(1.5)	(0.1)	(0.9)	(0.5)	0.2	(0.8)
110	*Vienna*	50.5	1.7	N	N	N	N	0	4.1	(1.5)	(0.1)	(0.9)	(0.5)	0.5	(0.8)
111	*West Indian*	51.8	7.0	N	N	N	N	0	4.8	(1.9)	N	N	N	N	(0.8)
112	*aerated (slimmers)*	48.8	2.1	N	N	N	N	0	3.9	1.8	0.2	1.1	0.5	0.2	0.9
113	**Wholemeal bread** *average*	39.8	1.8	N	N	N	N	0	7.4	(5.8)	(1.0)	(1.6)	(3.2)	0.6	(0.7)
114	*large*	40.0	1.9	N	N	N	N	0	7.3	5.8	1.0	1.6	3.2	0.7	(0.7)
115	*small, sliced*	38.5	1.8	N	N	N	N	0	7.4	(5.8)	(1.0)	(1.6)	(3.2)	0.6	(0.7)
116	*small, unsliced*	41.1	1.6	N	N	N	N	0	7.5	(5.8)	(1.0)	(1.6)	(3.2)	0.6	(0.7)
117	*toasted*	46.6	2.1	N	N	N	N	0	8.7	(5.9)	(1.1)	(1.5)	(3.3)	0.7	N

[a] High fibre types contain approximately 8.9g Englyst fibre per 100g

5.3

No. 11-	Food	Na	K	Ca	Mg	P	mg Fe	Cu	Zn	S	Cl	Mn	µg Se	I
99	**White bread** *average*	520	110	110	24	91	1.6	0.19	0.6	79	820	0.45	28	6
100	*large, crusty*	530	110	110	26	91	1.8	0.14	0.7	(79)	790	0.47	(28)	(6)
101	*large, tin*	510	110	110	24	90	1.6	0.14	0.6	(79)	810	0.45	(28)	(6)
102	*sliced*	530	99	100	20	79	1.4	0.13	0.5	(79)	830	0.41	(28)	(6)
103	*small, unwrapped*	510	110	110	25	110	1.7	0.28	0.7	(79)	790	0.47	(28)	(6)
104	*small, wrapped*	530	110	120	23	90	1.5	0.28	0.7	(79)	860	0.44	(28)	(6)
105	*fried*	550	100	100	21	82	1.5	0.14	0.5	(86)	860	0.42	(29)	(7)
106	*toasted*	650	120	120	24	96	1.7	0.16	0.6	(96)	1010	0.50	(34)	(7)
107	*French stick*	570	130	130	28	110	2.1	0.16	0.7	(79)	870	0.51	(28)	(6)
108	*Scottish batch unwrapped*	710	98	130	25	90	1.8	0.17	0.6	(79)	1250	0.48	(28)	(6)
109	*Scottish batch wrapped*	690	93	100	25	90	1.8	0.18	0.7	(79)	1100	0.47	(28)	(6)
110	*Vienna*	540	120	110	31	110	2.4	0.18	0.8	(79)	890	0.52	(28)	(6)
111	*West Indian*	450	94	150	39	95	2.6	0.38	1.0	(79)	630	0.66	(28)	(6)
112	*aerated (slimmers)*	560	110	130	21	93	1.7	0.14	0.6	N	950	0.42	N	N
113	**Wholemeal bread** *average*	550	230	54	76	200	2.7	0.26	1.8	81	880	1.90	35	Tr
114	*large*	540	220	65	71	200	2.6	0.25	1.7	(81)	630	1.80	(35)	Tr
115	*small, sliced*	570	230	63	78	200	2.9	0.29	1.9	(81)	1020	1.90	(35)	Tr
116	*small, unsliced*	550	230	33	80	210	2.6	0.23	1.8	(81)	970	1.90	(35)	Tr
117	*toasted*	640	270	63	89	230	3.2	0.30	2.1	(95)	1030	2.20	(41)	Tr

Breads continued

No. 11-	Food	Retinol equiv μg	Vitamin D μg	Vitamin E mg	Thiamin mg	Ribo-flavin mg	Niacin mg	Trypt 60 mg	Vitamin B6 mg	Vitamin B12 μg	Folate μg	Panto-thenate μg	Biotin μg	Vitamin C mg
99	**White bread** average	0	0	Tr	0.21	0.06	1.7	1.7	0.07	0	29	0.3	1	0
100	large, crusty	0	0	Tr	0.23	0.07	1.8	1.8	0.06	0	28	(0.3)	(1)	0
101	large, tin	0	0	Tr	0.22	0.07	1.7	1.7	0.07	0	17	(0.3)	(1)	0
102	sliced	0	0	Tr	0.20	0.05	1.5	1.6	0.07	0	17	(0.3)	(1)	0
103	small, unwrapped	0	0	Tr	0.20	0.03	1.8	1.8	0.07	0	25	(0.3)	(1)	0
104	small, wrapped	0	0	Tr	0.19	0.06	1.6	1.7	0.06	0	33	(0.3)	(1)	0
105	fried	0	0	Tr[a]	0.15	0.04	1.4	1.7	0.05	0	9	(0.2)	(1)	0
106	toasted	0	0	Tr	0.21	0.06	2.8	1.9	0.09	0	21	(0.4)	(1)	0
107	French stick	0	0	Tr	0.19	0.07	1.3	2.0	0.08	0	24	(0.3)	(1)	0
108	Scottish batch unwrapped	0	0	Tr	0.18	0.03	1.2	1.8	0.05	0	(21)	(0.3)	(1)	0
109	Scottish batch wrapped	0	0	Tr	0.21	0.04	1.3	1.7	0.06	0	21	(0.3)	(1)	0
110	Vienna	0	0	Tr	0.27	0.08	1.6	1.9	0.08	0	21	(0.3)	(1)	0
111	West Indian	0	0	Tr	0.16	0.03	1.2	1.7	(0.07)	0	25	(0.3)	(1)	0
112	aerated (slimmers)	0	0	Tr	0.31	0.07	1.5	1.9	0.07	0	42	N	N	0
113	**Wholemeal bread** average	0	0	0.20	0.34	0.09	4.1	1.8	0.12	0	39	0.6	6	0
114	large	0	0	(0.20)	0.26	0.08	4.0	1.8	0.11	0	33	(0.6)	(6)	0
115	small, sliced	0	0	(0.20)	0.38	0.09	4.1	1.9	0.12	0	43	(0.6)	(6)	0
116	small, unsliced	0	0	(0.20)	0.38	0.10	4.3	1.9	0.12	0	40	(0.6)	(6)	0
117	toasted	0	0	0.20	0.34	0.10	4.7	2.2	0.14	0	46	0.7	7	0

[a] The vitamin E content will depend on the fat used for frying

No. 11-	Food	Description and main data sources	Water g	Total nitrogen g	Protein g	Fat g	Carbohydrate g	Energy value kcal	Energy value kJ
118	**Brown rolls** *crusty*	12 samples of 6 rolls, different shops	30.5	1.81	10.3	2.8	50.4	255	1085
119	*soft*	14 samples of 6 rolls, different shops	31.6	1.75	10.0	3.8	51.8	268	1139
120	**Croissants**	Recipe	31.1	1.44	8.3	20.3	38.3	360	1505
121	**Hamburger buns**	5 packets of 6 buns including frozen	32.9	1.60	9.1	5.0	48.8	264	1121
122	**Morning rolls**	5 samples of 6 rolls, Scotland only	30.5	1.82	10.4	3.1	53.3	269	1144
123	**White rolls** *crusty*	14 samples of 6 rolls, different shops	26.4	1.92	10.9	2.3	57.6	280	1192
124	*soft*	14 samples of 6 rolls, different shops	32.7	1.61	9.2	4.2	51.6	268	1137
125	**Wholemeal rolls**	2 samples of 6 rolls, different shops	31.2	1.46	9.0	2.9	48.3	241	1025

No. 11-	Food	Starch	Total sugars	Individual sugars					Dietary fibre		Fibre fractions				
									Southgate method	Englyst method	Cellulose	Non-cellulosic polysaccharide		Lignin	Resistant starch
				Gluc	Fruct	Sucr	Malt	Lact				Soluble	Insoluble		
118	**Brown rolls** *crusty*	48.5	1.9	N	N	N	N	0	7.1	(3.5)	(0.5)	(1.1)	(1.8)	1.3	(0.8)
119	*soft*	49.3	2.5	N	N	N	N	0	6.4	(3.5)	(0.5)	(1.1)	(1.8)	0.9	(0.8)
120	**Croissants**	37.2	1.0	0.3	0.3	0.4	Tr	0	2.5	1.6	0.1	0.8	0.8	Tr	N
121	**Hamburger buns**	46.6	2.2	N	N	N	N	N	4.0	(1.5)	(0.1)	(0.9)	(0.5)	0.4	(0.8)
122	**Morning rolls**	51.1	2.2	N	N	N	N	0	4.1	(1.5)	(0.1)	(0.9)	(0.5)	0.1	(0.8)
123	**White rolls** *crusty*	55.4	2.2	N	N	N	N	0	4.3	(1.5)	(0.1)	(0.9)	(0.5)	0.6	(0.8)
124	*soft*	49.4	2.2	N	N	N	N	0	3.9	(1.5)	(0.1)	(0.9)	(0.5)	0.8	(0.8)
125	**Wholemeal rolls**	46.8	1.5	N	N	N	N	0	8.8	5.9	0.8	N	N	0.6	N

No. 11-	Food	Na	K	Ca	Mg	P	Fe	Cu	Zn	S	Cl	Mn	Se	I
							mg						µg	
118	Brown rolls *crusty*	570	200	100	65	190	3.2	0.34	1.5	170	1040	1.5 (1.40)	N	N
119	*soft*	560	190	110	59	170	3.4	0.23	1.5	+190	1130	1.40	N	N
120	Croissants	390	140	80	26	130	2.0	0.26	(0.9)	79	610	(0.35)	N	N
121	Hamburger buns	550	110	130	31	150	2.3	0.13	0.7	(130)	890	0.46	(28)	(19)
122	Morning rolls	530	120	120	36	120	2.5	0.26	0.9	(130)	810	0.64	(28)	(19)
123	White rolls *crusty*	640	130	140	34	120	2.1	0.19	0.9	(150)	1180	0.58	(28)	(19)
124	*soft*	560	120	120	26	100	2.2	0.13	0.7	(130)	930	0.46	(28)	19
125	Wholemeal rolls	460	230	55	69	170	3.5	0.25	1.6	(81)	850	1.60	(35)	Tr

No. 11-	Food	Retinol equiv µg	Vitamin D µg	Vitamin E mg	Thiamin mg	Ribo-flavin mg	Niacin mg	Trypt 60 mg	Vitamin B6 mg	Vitamin B12 µg	Folate µg	Panto-thenate µg	Biotin µg	Vitamin C mg
118	**Brown rolls** *crusty*	0	0	Tr	0.43	0.07	3.5	2.1	0.09	0	29	(0.3)	(3)	0
119	*soft*	0	0	Tr	0.41	0.08	3.4	2.0	0.09	0	29	(0.3)	(3)	0
120	**Croissants**	21	0.13	Tr	(0.18)	(0.16)	(2.0)	1.8	(0.11)	Tr	(73)	(0.5)	(9)	0
121	**Hamburger buns**	N	N	Tr	0.23	0.10	1.5	1.9	0.06	Tr	48	(0.3)	(1)	0
122	**Morning rolls**	0	0	Tr	0.23	0.07	2.0	2.1	0.08	0	26	(0.3)	(1)	0
123	**White rolls** *crusty*	0	0	Tr	0.26	0.05	2.1	2.2	0.04	0	32	(0.3)	(1)	0
124	*soft*	0	0	Tr	0.28	0.04	1.9	1.9	0.04	0	24	(0.3)	(1)	0
125	**Wholemeal rolls**	0	0	(0.20)	0.30	0.09	4.1	1.8	0.10	0	62	(0.6)	(6)	0

Breakfast cereals

11-126 to 11-140

Composition of food per 100g

No. 11-	Food	Description and main data sources	Water g	Total nitrogen g	Protein g	Fat g	Carbohydrate g	Energy value kcal	kJ
126	All-Bran	Analysis and manufacturer's data (Kelloggs)	3.0	2.40	15.1	3.4	43.0	252	1071
127	Bran Buds	Manufacturer's data (Kelloggs)	3.0	2.08	13.1	2.9	52.0	273	1162
128	Bran Flakes	Manufacturer's data (Kelloggs)	3.0	1.79	10.2	1.9	69.7	319	1359
129	Coco Pops	Manufacturer's data (Kelloggs)	3.0	0.90	5.3	1.4	93.9	386	1644
130	Corn Flakes	Analysis and manufacturer's data (Kelloggs)	3.0	1.26	7.9	0.6	84.9	355	1515
131	Crunchy Nut Corn Flakes	Manufacturer's data (Kelloggs)	3.0	1.18	7.4	4.0	88.6	398	1691
132	Farmhouse Bran	Manufacturer's data (Weetabix)	4.0	2.05	12.9	1.4	63.8	303	1292
133	Frosties	Manufacturer's data (Kelloggs)	3.0	0.85	5.3	0.4	95.4	382	1631
134	Fruit 'n Fibre	Manufacturer's data (Kelloggs)	5.7	1.42	8.1	5.1 ·	73.1	352	1496
135	Grapenuts	Manufacturer's data (General Foods)	3.5	1.80	10.5	0.5	79.9	346	1475
136	Honey Smacks	Manufacturer's data (Kelloggs)	3.0	1.23	7.0	1.0	89.1	371	1582
137	Muesli Swiss style[a]	Analysis and manufacturers' data (Kelloggs, Weetabix)	7.0	1.69	10.6	5.9	71.1	364	1546
138	with extra fruit	Manufacturer's data (Weetabix), as Alpen with tropical fruit	8.5	1.63	10.2	6.2	73.4	372	1577
139	with no added sugar	Analysis and manufacturer's data (Kelloggs)	6.7	1.67	10.5	8.1	67.1	366	1552
140	Nutri-Grain[b]	RYE & OATS with Hazel Nut variety, manufacturer's data (Kelloggs)	3.0	1.68	9.8	9.3	68.3	379	1603

[a]Muesli composition is very variable

[b]Nutri-grain - Wholewheat with Raisins variety contains 9.0 g protein, 1.2 g fat, 76.2 g carbohydrate, 11 g Southgate fibre, 7.0 g Englyst fibre, 333 kcal and 1417 kJ

Brown Rice and Rye variety contains 7.1g protein, 1.3 g, fat, 79.0 g carbohydrate, 10.2 g Southgate fibre, 4.2 g Englyst fibre, 336 kcal and 1433 kJ.

Breakfast Cereals

Carbohydrate fractions, g per 100g

No. 11-	Food	Starch	Total sugars	Individual sugars					Dietary fibre		Fibre fractions				
												Non-cellulosic polysaccharide			
				Gluc	Fruct	Sucr	Malt	Lact	Southgate method	Englyst method	Cellulose	Soluble	Insoluble	Lignin	Resistant starch
126	**All-Bran**	27.6	15.4	0.9	0.7	12.1	1.7	0	30.0	24.5	4.5	4.1	15.9	1.5	0.1
127	**Bran Buds**	26.3	25.7	1.3	0.6	21.6	2.2	0	28.2	20.0	N	N	N	N	N
128	**Bran Flakes**	50.7	19.0	1.7	1.9	14.9	0.5	0	17.3	11.3	2.1	3.0	6.2	N	0.7
129	**Coco Pops**	55.7	38.2	1.2	1.0	35.5	0	0.5	1.1	0.6	0.2	0.2	0.2	N	N
130	**Corn Flakes**	77.7	7.2	1.4	1.4	4.0	0.4	0	3.4	0.9	0.3	0.4	0.2	N	2.9
131	**Crunchy Nut Corn Flakes**	52.3	36.3	N	N	31.8	N	N	1.6	0.8	0.3	0.3	0.2	N	N
132	**Farmhouse Bran**	40.7	23.1	3.0	1.4	16.8	1.9	0	18.0	N	3.6	N	N	2.2	0.2
133	**Frosties**	53.9	41.5	N	N	38.9	N	0	1.2	0.5	0.2	0.2	0.1	N	1.2
134	**Fruit 'n Fibre**	46.4	26.7	N	N	14.1	N	0	10.1	7.0	0.9	2.6	3.5	N	0.7
135	**Grapenuts**	67.8	12.1	(1.0)	(4.3)	(0)	(6.8)	0	6.2	N	N	N	N	0.6	0.6
136	**Honey Smacks**	49.9	39.2	N	N	34.3	N	0	4.6	4.7	0.7	1.6	2.4	N	0.1
137	**Muesli** Swiss style	47.4	23.7	4.6	4.7	10.9	0.2	4.5	8.1	6.1	0.9	1.8	3.5	N	N
138	with extra fruit	51.1	22.3	3.9	3.9	10.0	0.1	4.5	5.4	N	0.8	N	N	0.4	N
139	with no added sugar	51.8	15.3	6.3	7.1	1.4	0.6	Tr	11.1	7.7	N	N	N	N	N
140	**Nutri-Grain**	59.9	8.4	N	N	2.0	N	0	12.4	8.9	N	N	N	N	N

Inorganic constituents per 100g

No. 11-	Food	Na	K	Ca	Mg	P	Fe	Cu	Zn	S	Cl	Mn	Se	I
		mg											µg	
126	All-Bran	1480	900	69	370	620	12.0	0.44	8.4	N	2390	N	N	N
127	Bran Buds	510	920	56	N	570	12.0	0.53	N	N	800	N	N	N
128	Bran Flakes	910	540	50	N	370	40.0	0.35	N	N	1460	N	N	N
129	Coco Pops	880	190	35	N	120	6.7	0.20	N	N	1400	N	N	N
130	Corn Flakes	1110	100	15	14	38	6.7	0.03	0.3	N	1820	0.08	2	10
131	Crunchy Nut Corn Flakes	770	150	18	N	43	6.7	0.08	N	N	1210	N	N	N
132	Farmhouse Bran	870	730	70	180	620	44.0	1.00	4.0	N	1340	N	N	N
133	Frosties	740	63	11	N	25	6.7	Tr	N	N	1150	N	N	N
134	Fruit 'n Fibre	560	450	51	N	200	6.7	0.24	N	N	970	N	N	N
135	Grapenuts	590	310	37	95	250	9.5	0.45	4.2	N	1080	N	N	N
136	Honey Smacks	320	250	100	N	150	6.7	0.21	N	N	610	N	N	N
137	Muesli *Swiss style*	380	440	120	85	280	5.6	0.10	2.5	N	790	N	N	N
138	*with extra fruit*	120	420	66	85	290	4.0	Tr	2.0	N	190	N	N	N
139	*with no added sugar*	29	510	49	96	310	3.4	0.36	2.1	N	10	2.60	N	N
140	Nutri-Grain	500	370	45	N	330	6.7	0.07	N	N	1030	N	N	N

Breakfast cereals

No. 11-	Food	Retinol equiv µg	Vitamin D µg	Vitamin E mg	Thiamin mg	Riboflavin mg	Niacin mg	Trypt 60 mg	Vitamin B6 mg	Vitamin B12 µg	Folate µg	Pantothenate µg	Biotin µg	Vitamin C mg
126	All-Bran	0	2.80	2.20	1.00	1.50	16.0	3.2	1.80	2	250	1.7	25	0
127	Bran Buds	0	2.80	N	1.00	1.50	16.0	2.8	1.80	2	250	2.5	21	0
128	Bran Flakes	0	2.80	N	1.00	1.50	16.0	2.4	1.80	2	250	0.9	11	35
129	Coco Pops	0	2.80	N	1.00	1.50	16.0	1.2	1.80	2	250	N	N	0
130	Corn Flakes	0	2.80	0.40	1.00	1.50	16.0	0.9	1.80	2	250	0.3	2	0
131	Crunchy Nut Corn Flakes	0	2.80	N	1.00	1.50	16.0	0.8	1.80	2	250	N	N	N
132	Farmhouse Bran	0	0	0	1.60	1.80	19.0	2.7	0	0	N	N	N	0
133	Frosties	0	2.80	N	1.00	1.50	16.0	0.6	1.80	2	250	(0.3)	(1)	0
134	Fruit 'n Fibre	Tr	2.80	N	1.00	1.50	16.0	1.7	1.80	2	250	N	N	N
135	Grapenuts	1320	4.40	1.70	1.30	1.50	17.6	2.6	1.80	5	350	N	N	0
136	Honey Smacks	0	2.80	N	1.00	1.50	16.0	1.6	1.80	2	250	N	N	0
137	Muesli *Swiss style*	Tr	0	3.20	0.50	0.70	6.5	2.3	1.60	0	(140)	1.2	15	Tr
138	*with extra fruit*	N	0	N	0.10	0.50	5.0	2.2	Tr	0	N	N	N	2
139	*with no added sugar*	Tr	0	(3.20)	0.30	0.30	5.3	2.2	N	0	N	N	N	Tr
140	Nutri-Grain	0	2.80	N	1.00	1.50	16.0	1.8	1.80	2	250	N	N	N

Breakfast cereals *continued*

Composition of food per 100g

No. 11-	Food	Description and main data sources	Water g	Total nitrogen g	Protein g	Fat g	Carbohydrate g	Energy value kcal	kJ
141	**Porridge** *made with milk*	Recipe	74.8	0.77	4.8	5.1	13.7	116	488
142	*made with milk and water*	Recipe, ref. 7	81.1	0.52	3.2	3.1	11.3	83	348
143	*made with water*	Recipe, ref. 7	87.4	0.26	1.5	1.1	9.0	49	209
144	**Puffed Wheat**	Analytical and literature sources	2.5	2.44	14.2	1.3	67.3	321	1366
145	**Ready Brek**	6 packets of the same brand (Lyons)	6.3	2.12	12.4	8.7	69.5	389	1645
146	**Rice Krispies**	Analysis and manufacturer's data (Kelloggs)	3.8	1.03	6.1	0.9	89.7	369	1572
147	**Ricicles**	Manufacturer's data (Kelloggs)	3.0	0.72	4.3	0.5	96.3	383	1632
148	**Shredded Wheat**	6 packets of the same brand (Nabisco)	7.6	1.81	10.6	3.0	68.3	325	1384
149	**Shreddies**	10 samples (Nabisco)	4.0	1.72	10.0	1.5	74.1	331	1411
150	**Special K**	Analysis and manufacturer's data (Kelloggs)	2.7	2.46	15.3	1.0	82.5	380	1617
151	**Start**	Manufacturer's data (Kelloggs)	3.0	1.26	7.9	1.7	82.0	354	1509
152	**Sugar Puffs**	6 packets of the same brand (Quaker)	1.8	1.01	5.9	0.8	84.5	324	1381
153	**Sultana Bran**	Manufacturer's data (Kelloggs)	7.0	1.49	8.5	1.6	67.7	302	1287
154	**Weetabix**	Analysis and manufacturer's data (Weetabix)	5.8	1.80	10.7	2.0	70.3 *74.9*	363	1454
155	**Weetaflake**	Analysis and manufacturer's data (Weetabix)	5.8	1.84	10.7	2.0	74.9	342	1454
156	**Weetaflake 'n' raisin**	Manufacturer's data (Weetabix)	8.0	1.37	8.2	0.6	79.8	337	1438
157	**Weetos**	Manufacturer's data (Weetabix)	3.0	1.40	8.2	0.7	80.5	341	1453

Breakfast cereals *continued*

Carbohydrate fractions, g per 100g

No. 11-	Food	Starch	Total sugars	Individual sugars					Dietary fibre		Fibre fractions				
				Gluc	Fruct	Sucr	Malt	Lact	Southgate method	Englyst method	Cellulose	Non-cellulosic polysaccharide Soluble	Insoluble	Lignin	Resistant starch
141	**Porridge** *made with milk*	9.0	4.7	Tr	Tr	Tr	Tr	4.7	0.8	0.8	0.1	0.5	0.3	N	N
142	*made with milk and water*	9.0	2.4	Tr	Tr	Tr	Tr	2.4	0.8	0.8	0.1	0.5	0.3	N	N
143	*made with water*	9.0	Tr	Tr	Tr	Tr	Tr	Tr	0.8	0.8	0.1	0.5	0.3	N	N
144	**Puffed Wheat**	67.0	0.3	Tr	0.1	0.2	0	0	8.8	5.6	Tr	N	N	1.2	0.7
145	**Ready Brek**	67.7	1.7[a]	0.1	Tr	1.1	0.6	0	6.8	7.2	0.8	3.1	3.3	N	0.3
146	**Rice Krispies**	79.1	10.6	1.0	0.7	8.8	0.1	0	1.1	0.5	0.2	0.1	0.2	0.1	0.2
147	**Ricicles**	56.5	39.8	N	N	38.7	N	0	0.9	0.3	0.2	Tr	0.1	N	0.2
148	**Shredded Wheat**	67.5	0.8	Tr	Tr	0.8	0	0	10.1	9.8	1.6	2.0	6.2	1.1	0.8
149	**Shreddies**	63.9	10.2	3.2	3.7	1.5	1.9	0	10.9	9.5	1.5	N	N	0.7	1.0
150	**Special K**	64.5	18.0	0.6	0.4	15.2	1.2	0.6	2.7	2.0	0.3	0.8	0.9	N	0.3
151	**Start**	50.5	31.5	2.6	1.6	27.3	Tr	0	9.3	5.7	0.9	1.8	3.0	N	0.1
152	**Sugar Puffs**	28.0	56.5	4.5	2.0	45.6	4.4	0	4.8	3.2	0.6	1.6	1.0	0.7	0.3
153	**Sultana Bran**	34.9	32.9	N	N	13.1	N	0	15.5	10.0	1.8	4.3	3.9	N	0.5
154	**Weetabix**	71.9 *(68.5)*	6.4	1.8	0.8	2.9	0.8	0	8.5 11.6	9.7	1.5	3.1	5.1	1.1	0.2
155	**Weetaflake**	68.5	6.4	1.8	0.8	2.9	0.8	0	11.6	(9.7)	(1.5)	(3.1)	(5.1)	(1.1)	(0.2)
156	**Weetaflake 'n' raisin**	55.2	24.6	11.3	11.8	0.8	0.6	0	7.2	N	0.9	N	N	1.4	N
157	**Weetos**	51.1	29.4	Tr	Tr	28.9	Tr	0.5	11.7	N	2.6	N	N	1.6	N

[a]Flavoured instant oat varieties contain approximately 8.5g total sugars per 100g

45

Breakfast cereals *continued*

Inorganic constituents per 100g

No. 11-	Food	Na	K	Ca	Mg	P	Fe	Cu	Zn	S	Cl	Mn	Se	I
						mg							µg	
141	**Porridge** *made with milk*	620	190	120	29	140	0.6	0.03	0.8	51	970	0.46	Tr	N
142	*made with milk and water*	590	120	66	23	94	0.5	0.03	0.6	35	920	0.46	Tr	N
143	*made with water*	560	46	7	18	47	0.5	0.03	0.4	20	870	0.46	Tr	N
144	**Puffed Wheat**	4	390	26	140	350	4.6	0.56	2.8	N	50	N	N	N
145	**Ready Brek**	23	390	65	120	420	4.8	0.41	2.7	N	66	N	N	N
146	**Rice Krispies**	1260	150	20	50	130	6.7	0.10	1.1	N	1980	1.00	N	N
147	**Ricicles**	880	96	20	N	89	6.7	0.13	N	N	1400	N	N	N
148	**Shredded Wheat**	8	330	38	130	340	4.2	0.40	2.3	N	53	N	N	N
149	**Shreddies**	550	210	40	88	320	2.8	0.44	2.5	100	220	2.30	N	N
150	**Special K**	1150	230	70	52	240	13.3	0.13	1.9	N	1640	N	N	N
151	**Start**	410	300	68	N	190	15.0	0.13	18.7	N	910	N	N	N
152	**Sugar Puffs**	9	160	14	55	140	2.1	0.23	1.5	N	41	N	N	N
153	**Sultana Bran**	610	660	51	N	310	30.0	0.13	N	N	1090	N	N	N
154	**Weetabix**	370	370	35	120	290	6.0	0.54	2.0	N	570	N	N	N
155	**Weetaflake**	370	370	35	120	290	6.0	(0.54)	(2.0)	N	(570)	N	N	N
156	**Weetaflake 'n' raisin**	200	550	77	70	230	7.0	N	N	N	310	N	N	N
157	**Weetos**	830	490	65	120	290	10.0	Tr	2.3	N	N	N	N	N

46

Breakfast cereals *continued*

No. 11-	Food	Retinol equiv µg	Vitamin D µg	Vitamin E mg	Thiamin mg	Ribo-flavin mg	Niacin mg	Trypt 60 mg	Vitamin B6 mg	Vitamin B12 µg	Folate µg	Panto-thenate µg	Biotin µg	Vitamin C mg
141	**Porridge** *made with milk*	56	0.03	0.29	0.10	0.17	0.2	1.1	0.06	0	7	0.4	3	1
142	*made with milk and water*	28	0.02	0.25	0.08	0.09	0.2	0.7	0.04	0	5	0.2	2	Tr
143	*made with water*	0	0	0.21	0.06	0.01	0.1	0.3	0.01	0	4	0.1	2	0
144	**Puffed Wheat**	0	0	2.00	Tr	0.06	5.2	2.9	0.14	0	19	0.5	7	0
145	**Ready Brek**	0	0	1.20	1.50	0.09	9.4	2.8	1.50	0	53	1.3	23	0
146	**Rice Krispies**	0	2.80	0.60	1.00	1.50	16.0	1.4	1.80	2	250	0.7	2	0
147	**Ricicles**	0	2.80	N	1.00	1.50	16.0	1.0	1.80	2	250	(0.4)	(1)	0
148	**Shredded Wheat**	0	0	1.20	0.27	0.05	4.5	2.1	0.24	0	29	0.8	9	0
149	**Shreddies**	0	0	N	1.20	2.20	21.1	2.0	0.64	1	28	0.8	7	0
150	**Special K**	0	2.80	0.55	1.20	1.70	18.3	2.8	2.20	2	300	0.5	3	0
151	**Start**	0	4.20	18.70	1.50	2.20	24.0	1.1	2.70	3	400	N	N	38
152	**Sugar Puffs**	0	0	0.34	Tr	0.03	2.5	1.2	0.05	0	12	N	N	0
153	**Sultana Bran**	Tr	2.80	N	1.00	1.50	16.0	2.0	1.80	2	250	N	N	N
154	**Weetabix**	0	0	1.03	0.70	1.00	10.0	2.1	0.22	0	50	0.7	8	0
155	**Weetaflake**	0	0	1.03	0.20	1.00	10.0	2.1	0.22	0	50	(0.7)	(8)	0
156	**Weetaflake 'n' raisin**	Tr	0	N	0.60	1.30	9.6	1.6	N	0	N	N	N	1
157	**Weetos**	7	2.80	Tr	1.00	1.50	16.0	1.6	Tr	3	2	N	N	30

Infant foods

Composition of food per 100g

No. 11-	Food	Description and main data sources	Water g	Total nitrogen g	Protein g	Fat g	Carbohydrate g	Energy value kcal	kJ
158	**Baby cereals,** rice-based	10 samples, 4 brands	4.9	1.84	10.9	4.8	79.6	386	1639
159	various cereal-based	17 samples, 10 varieties, e.g. oat based, mixed cereal and fruit, etc.	5.4	1.96	11.4	9.1	72.4	399	1689
160	wheat-based	10 samples, 4 brands	4.2	3.22	18.8	5.8	70.3	391	1658
161	**Rusks,** plain	10 samples, 4 brands	5.0	1.11	6.5	7.9	82.8	408	1729
162	low sugar	10 samples, 5 brands, plain and flavoured	5.5	1.47	8.6	9.7	77.8	414	1751
163	flavoured	15 samples, 2 brands, assorted flavours	7.1	1.16	6.8	9.0	78.1	401	1698
164	wholemeal	8 samples, 1 brand (Farleys)	4.4	1.42	8.3	10.1	76.5	411	1739

Infant foods

Carbohydrate fractions, g per 100g

No. 11-	Food	Starch	Total sugars	Individual sugars					Dietary fibre		Fibre fractions				
				Gluc	Fruct	Sucr	Malt	Lact	Southgate method	Englyst method	Cellulose	Non-cellulosic polysaccharide Soluble	Insoluble	Lignin	Resistant starch
158	**Baby cereals**, rice-based	73.3	6.3	0.5	0.1	3.3	0.8	1.6	4.0	N	N	N	N	N	N
159	various cereal-based	48.9	23.5	3.2	1.8	7.6	1.8	9.1	4.1	N	N	N	N	N	N
160	wheat-based	60.5	9.7	0.3	0.2	2.9	2.1	4.2	4.0	N	N	N	N	N	N
161	**Rusks**, plain	50.8	32.0	0.2	0.1	30.9	0.5	0.3	2.6	N	N	N	N	N	N
162	low sugar	54.3	23.5	3.9	0.5	15.0	1.6	2.5	2.7	N	N	N	N	N	N
163	flavoured	55.6	22.5	1.6	0.8	17.0	2.0	1.1	3.7	N	N	N	'N	N	N
164	wholemeal	51.4	25.1	3.4	0.3	19.2	0.5	1.7	4.5	N	N	N	N	N	N

No. 11-	Food	Na	K	Ca	Mg	P	Fe	Cu	Zn	S	Cl	Mn	Se	I
							mg						µg	
158	**Baby cereals**, rice-based	35	340	870	72	370	20.0	0.43	2.3	N	54	1.60	N	N
159	various cereal-based	110	970	710	71	570	15.0	0.40	2.3	N	170	1.90	N	N
160	wheat-based	68	1040	830	84	420	18.0	0.50	2.7	N	110	1.50	N	N
161	**Rusks**, plain	66	140	520	23	190	28.0	0.21	1.1	N	100	0.70	N	N
162	low sugar	110	200	530	31	160	26.0	0.24	1.0	N	160	0.70	N	N
163	flavoured	87	200	440	37	170	20.0	0.29	1.1	N	130	0.73	N	N
164	wholemeal	23	250	610	59	260	23.0	0.35	1.7	N	35	1.70	N	N

No. 11-	Food	Retinol equiv μg	Vitamin D μg	Vitamin E mg	Thiamin mg	Ribo-flavin mg	Niacin mg	Trypt 60 mg	Vitamin B6 mg	Vitamin B12 μg	Folate μg	Panto-thenate μg	Biotin μg	Vitamin C mg
158	**Baby cereals**, rice-based	N	N	N	1.60	1.20	23.0	2.5	0.40	1	82	1.9	6	100
159	various cereal-based	N	N	N	1.10	1.10	9.0	2.3	0.49	2	92	4.9	16	71
160	wheat-based	N	N	N	2.00	1.70	27.0	3.8	0.40	1	69	1.8	9	130
161	**Rusks**, plain	N	N	N	0.60	1.30	12.0	1.3	0.09	Tr	10	0.2	1	Tr
162	low sugar	N	N	N	0.70	1.40	12.0	1.7	0.14	Tr	14	0.5	3	Tr
163	flavoured	N	N	N	0.90	0.90	7.0	1.3	0.70	Tr	62	1.6	4	Tr
164	wholemeal	N	N	N	0.70	1.00	14.0	1.7	0.32	Tr	24	0.5	4	Tr

51

No. 11-	Food	Description and main data sources	Water g	Total nitrogen g	Protein g	Fat g	Carbohydrate g	Energy value kcal	kJ
165	**Brandy snaps**	Recipe	13.7	0.44	2.5	20.3	64.0	437	1834
166	**Chocolate biscuits,**								
	full coated	7 different kinds	2.2	1.00	5.7	27.6	67.4	524	2197
167	**Cream crackers**	6 packets	4.3	1.66	9.5	16.3	68.3	440	1857
168	**Crispbread**, rye	Analytical and literature sources	6.4	1.61	9.4	2.1	70.6	321	1367
169	**Digestive biscuits,**								
	chocolate	10 packets, 5 plain chocolate, 5 milk chocolate	2.5	1.17	6.8	24.1	66.5	493	2071
170	plain	10 samples, 3 brands	2.5	1.10	6.3	20.9	68.6	471	1978
171	**Flapjacks**	Recipe	6.4	0.77	4.5	26.6	60.4	484	2028
172	**Gingernut biscuits**	10 packets, 6 brands	3.4	0.98	5.6	15.2	79.1	456	1923
173	homemade	Recipe	6.6	0.77	4.4	19.4	68.2	448	1884
174	**Homemade biscuits**								
	creaming method	Recipe	9.0	1.07	6.2	21.9	64.3	463	1943
175	*rubbing-in method*	Recipe	9.2	1.11	6.5	23.1	61.7	466	1952
176	wholemeal	Recipe	9.8	2.26	13.4	8.9	60.0	358	1515
177	**Jaffa cakes**	Recipe	18.0	0.58	3.5	10.5	67.8	363	1532
178	**Matzos**	6 packets, Rakusens, Superfine, tea	6.7	1.85	10.5	1.9	86.6	384	1634
179	**Melting moments**	Recipe	4.7	0.57	3.3	36.5	55.2	549	2290
180	**Oatcakes** *homemade*	Recipe	9.2	1.84	10.8	18.3	63.2	445	1872
181	*retail*	6 packets, 4 brands	5.5	1.71	10.0	18.3	63.0	441	1855
182	**Sandwich biscuits**	10 packets, custard creams and similar types	2.6	0.87	5.0	25.9	69.2	513	2151

Carbohydrate fractions, g per 100g

No. 11-	Food	Starch	Total sugars	Gluc	Fruct	Sucr	Malt	Lact	Southgate method	Englyst method	Cellulose	Soluble	Insoluble	Lignin	Resistant starch
						Individual sugars			Dietary fibre			Non-cellulosic polysaccharide	Fibre fractions		
165	**Brandy snaps**	18.9	45.1	0.2	0.2	44.7	Tr	0	0.9	0.8	Tr	0.4	0.4	Tr	N
166	**Chocolate biscuits**, full coated	24.0	43.4	0	0	38.2	0	5.2	2.9	2.1	0.5	N	N	1.3	Tr
167	**Cream crackers**	68.3	Tr	Tr	Tr	Tr	Tr	0	6.1	2.2	0.1	1.4	0.7	1.3	0.6
168	**Crispbread**, rye	67.4	3.2	0.5	0.9	1.3	0.5	0	11.6[a]	11.7[ab]	0.9	3.9	6.9	1.1	1.8
169	**Digestive biscuits**, chocolate	38.0	28.5	N	N	26.0	0	2.5	3.1	2.2	0.3	1.1	0.8	0.5	0.2
170	plain	55.0	13.6	0.3	0.3	13.0	0	0	4.6	2.2	0.3	1.1	0.9	0.5	0.3
171	**Flapjacks**	25.0	35.5	Tr	Tr	35.2	0.1	0	2.6	2.7	0.2	1.6	0.9	0.6	N
172	**Gingernut biscuits**	43.3	35.8	2.1	0.9	32.8	0	0	1.8	1.4	0.1	0.9	0.4	0.2	0.2
173	homemade	35.4	32.8	0.4	0.4	32.0	0	0	2.0	1.4	0.1	0.7	0.7	N	N
174	**Homemade biscuits** _creaming method_	37.6	26.7	0.3	0.3	26.0	0.1	0	1.8	1.5	0.1	0.7	0.7	Tr	N
175	_rubbing-in method_	40.3	21.5	0.3	0.3	20.8	0.1	0	1.9	1.6	0.1	0.8	0.8	Tr	N
176	wholemeal	55.3	4.7	0.5	0.5	0.9	0	2.8	7.7	8.1	1.3	1.8	5.0	0.3	N
177	**Jaffa cakes**	10.6	57.2	11.2	6.2	35.7	3.4	0.7	N	N	N	N	N	Tr	N
178	**Matzos**	82.4	4.2	0	0	0	4.2	0	3.5	3.0	0.3	1.6	1.1	0.4	0.8
179	**Melting moments**	35.6	19.6	2.8	1.6	12.8	2.4	0	1.2	1.2	Tr	0.5	0.5	Tr	N
180	**Oatcakes** _homemade_	63.2	Tr	Tr	Tr	Tr	Tr	0	5.5	5.9	0.4	3.5	2.0	N	N
181	_retail_	59.9	3.1	Tr	Tr	2.0	1.1	0	3.6	N	N	N	N	0.4	N
182	**Sandwich biscuits**	39.0	30.2	1.0	0.9	27.8	0	0.5	1.1	N	N	N	N	N	N

[a] Cracotte type crispbread contains 9.2g Southgate fibre and 3.5g Englyst fibre per 100g
[b] High fibre varieties contain approximately 17.9g Englyst fibre per 100g

Biscuits

Inorganic constituents per 100g

No. 11-	Food	Na	K	Ca	Mg	P	Fe	Cu	Zn	S	Cl	Mn	Se	I
							mg						µg	
165	**Brandy snaps**	250	110	45	9	39	1.0	0.07	0.3	N	350	N	1	12
166	**Chocolate biscuits**, full coated	160	230	110	42	130	1.7	0.25	0.8	N	250	N	N	N
167	**Cream crackers**	610	120	110	25	110	1.7	(0.20)	(0.7)	87	830	(0.60)	(4)	(13)
168	**Crispbread**, rye	220[a]	500	45[a]	100	310	3.5[a]	0.38	3.0[a]	140	370	3.50	(3)	15
169	**Digestive biscuits**, chocolate	450	210	84	41	130	2.1	0.24	1.0	N	410	N	N	N
170	plain	600	170	92	23	88	3.2	0.28	0.5	110	540	0.45	N	N
171	**Flapjacks**	280	200	37	48	160	2.1	0.23	1.5	97	370	1.97	1	N
172	**Gingernut biscuits**	330	220	130	25	87	4.0	0.16	0.5	N	320	(0.87)	N	N
173	homemade	830	170	170	17	220	1.9	0.13	0.6	N	580	0.87	2	N
174	**Homemade biscuits** *creaming*													
	method	220	92	78	12	82	1.3	0.10	0.6	N	360	0.30	4	18
175	*rubbing-in method*	410	96	91	13	150	1.4	0.12	0.6	N	520	0.32	4	18
176	wholemeal	470	390	110	120	340	3.5	0.41	2.8	N	760	2.81	47	N
177	**Jaffa cakes**	130	170	55	34	130	1.5	0.30	0.3	N	170	N	N	48
178	**Matzos**	17	150	32	20	100	1.5	0.16	0.7	N	80	N	N	N
179	**Melting moments**	370	62	57	8	47	1.1	0.09	0.4	N	580	N	1	414
180	**Oatcakes** *homemade*	510	320	48	98	330	3.6	0.20	2.9	142	640	3.21	3	N
181	*retail*	1230	340	54	100	420	4.5	0.37	2.3	(140)	1290	(3.20)	N	N
182	**Sandwich biscuits**	220	120	100	13	82	1.6	0.07	0.5	N	290	N	N	N

[a]Cracotte type crispbread contains 640mg Na, 80mg Ca, 2.1mg Fe and 0.6mg Zn per 100g

Biscuits

No. 11-	Food	Retinol equiv µg	Vitamin D µg	Vitamin E mg	Thiamin mg	Ribo-flavin mg	Niacin mg	Trypt 60 mg	Vitamin B6 mg	Vitamin B12 µg	Folate µg	Panto-thenate µg	Biotin µg	Vitamin C mg
165	Brandy snaps	210	0.18	0.56	0.06	0.01	0.4	0.5	0.03	0	3	0.1	Tr	0
166	Chocolate biscuits, full coated	Tr	0	1.43	0.03	0.13	0.5	1.2	0.04	0	N	N	N	0
167	Cream crackers	0	0	(1.30)	(0.23)	(0.05)	(1.7)	1.9	(0.12)	0	(22)	(0.3)	(2)	0
168	Crispbread, rye	0	0	0.50	0.28	0.14	1.1	1.8	0.29	0	35	(1.1)	(7)	0
169	Digestive biscuits, chocolate	Tr	0	1.10	0.08	0.11	1.3	1.4	0.08	0	N	N	N	0
170	plain	0	0	N	0.14	0.11	1.1	1.3	0.09	0	13	N	N	0
171	Flapjacks	230	2.27	2.85	0.26	0.03	0.3	1.0	0.09	0	11	0.3	8	0
172	Gingernut biscuits	N	0	1.50	0.10	0.03	0.9	1.1	0.07	0	(4)	(0.1)	(1)	0
173	homemade	190	1.84	2.00	0.10	0.01	0.7	0.8	0.05	0	4	0.1	1	0
174	Homemade biscuits *creaming method*	220	2.11	2.32	0.12	0.06	0.8	1.4	0.07	Tr	9	0.3	4	0
175	*rubbing-in method*	230	2.22	2.44	0.13	0.06	0.9	1.4	0.07	Tr	9	0.3	3	0
176	*wholemeal*	79	0.46	1.76	0.34	0.16	4.9	2.7	0.36	Tr	27	0.7	7	1
177	Jaffa cakes	14	0.12	0.81	0.05	0.05	0.3	0.7	0.03	0	5	0.2	3	2
178	Matzos	0	0	Tr	0.11	0.03	0.9	2.2	0.06	0	N	N	N	0
179	Melting moments	350	3.53	3.66	0.08	0.01	0.5	0.6	0.04	0	4	0.1	0	0
180	Oatcakes *homemade*	0	0	1.48	0.33	0.07	0.8	2.4	0.08	0	26	0.7	17	0
181	*retail*	0	0	2.14	0.32	0.09	0.7	2.3	0.10	0	(26)	(1.0)	(17)	0
182	Sandwich biscuits	0	0	3.40	0.14	0.13	1.1	1.0	0.04	0	N	N	N	0

Biscuits *continued*

11-183 to 11-188

Composition of food per 100g

No. 11-	Food	Description and main data sources	Water g	Total nitrogen g	Protein g	Fat g	Carbohydrate g	Energy value kcal	kJ
183	**Semi-sweet biscuits**	10 packets, Osborne, Rich Tea, Marie	2.5	1.18	6.7	16.6	74.8	457	1925
184	**Short-sweet biscuits**	10 packets, shortcake and Lincoln	2.6	1.08	6.2	23.4	62.2	469	1966
185	**Shortbread**	Recipe	5.8	1.04	5.9	26.1	63.9	498	2087
186	**Wafer biscuits,** filled	9 packets, assorted	2.3	0.82	4.7	29.9	66.0	535	2242
187	**Water biscuits**	3 brands	4.5	1.90	10.8	12.5	75.8	440	1859
188	**Wholemeal crackers**	Farmhouse-type, recipe	4.4	1.76	10.1	11.3	72.1	413	1744

Biscuits *continued*

Carbohydrate fractions, g per 100g

No. 11-	Food	Starch	Total sugars	Individual sugars					Dietary fibre		Fibre fractions				
				Gluc	Fruct	Sucr	Malt	Lact	Southgate method	Englyst method	Cellulose	Non-cellulosic polysaccharide Soluble	Insoluble	Lignin	Resistant starch
183	Semi-sweet biscuits	52.5	22.3	0	0	19.1	3.2	0	2.1	1.7	0.1	1.1	0.5	0.2	0.3
184	Short-sweet biscuits	38.1	24.1	1.4	0	22.7	0	0	1.5	1.5	0.8	0.6	0.1	0.1	N
185	Shortbread	46.9	17.1	0.3	0.3	16.3	0.1	0	2.2	1.9	0.1	0.9	0.9	Tr	N
186	Wafer biscuits, filled	21.3	44.7	1.4	0	42.9	0	0.4	1.4	N	N	N	N	0.1	N
187	Water biscuits	73.5	2.3	0	0	0	(2.3)	0	(6.1)	3.1	0.2	1.8	1.1	(1.3)	0.5
188	Wholemeal crackers	70.5	1.6	0.5	0.5	0.4	0.1	0	4.8	4.4	0.4	1.6	2.5	0.1	N

Biscuits *continued*

No. 11-	Food	Na	K	Ca	Mg	P	mg Fe	Cu	Zn	S	Cl	Mn	µg Se	I
183	**Semi-sweet biscuits**	410	140	120	17	84	2.1	0.08	0.6	N	520	N	N	N
184	**Short-sweet biscuits**	360	110	87	15	85	1.8	0.11	0.6	N	490	N	N	N
185	**Shortbread**	230	97	91	13	75	1.3	0.10	0.4	N	460	0.37	2	18
186	**Wafer biscuits**, filled	70	160	73	22	83	1.6	0.16	0.6	N	150	N	N	N
187	**Water biscuits**	470	140	120	19	87	1.6	0.08	(0.7)	100	680	N	N	N
188	**Wholemeal crackers**	700	200	110	49	170	2.5	0.25	1.2	N	1040	1.20	N	N

Biscuits *continued*

No. 11-	Food	Retinol equiv µg	Vitamin D µg	Vitamin E mg	Thiamin mg	Ribo-flavin mg	Niacin mg	Trypt 60 mg	Vitamin B6 mg	Vitamin B12 µg	Folate µg	Panto-thenate µg	Biotin µg	Vitamin C mg
183	**Semi-sweet biscuits**	0	0	1.40	0.13	0.08	1.5	1.4	0.06	0	(13)	N	N	0
184	**Short-sweet biscuits**	0	0	1.30	0.16	0.04	0.9	1.3	0.05	0	(13)	N	N	0
185	**Shortbread**	270	0.23	0.80	0.14	0.02	1.0	1.2	0.07	0	7	0.1	1	0
186	**Wafer biscuits**, filled	0	0	1.90	0.09	0.08	0.5	1.0	0.03	0	N	N	N	0
187	**Water biscuits**	0	0	N	(0.11)	(0.03)	(0.9)	2.2	(0.06)	0	N	N	N	0
188	**Wholemeal crackers**	0	0	1.53	0.26	0.06	2.7	2.0	0.18	0	26	0.4	3	0

Cakes

Composition of food per 100g

No. 11-	Food	Description and main data sources	Water g	Total nitrogen g	Protein g	Fat g	Carbohydrate g	Energy value kcal	kJ
189	**All-Bran loaf**	Recipe. Ref. 7	26.6	0.84	5.2	1.6	58.4	254	1081
190	**Battenburg cake**	Recipe. Ref. 7	25.3	1.02	5.9	17.5	50.0	370	1551
191	**Cake mix**	10 samples, 3 brands	5.3	0.82	4.7	2.5	77.2	331	1408
192	*made up*	Recipe; made as packet directions	31.5	0.89	5.3	3.3	52.4	248	1052
193	**Cherry cake**	Recipe	14.4	0.85	5.1	15.8	61.7	394	1657
194	**Chinese cakes and biscuits**	4 assorted samples	19.8	1.18	6.7	21.5	51.9	415	1740
195	**Chocolate cake**	Recipe. Ref. 7	12.7	1.27	7.4	26.4	50.4	456	1908
196	*with butter icing*	Recipe. Ref. 7	11.6	0.98	5.7	29.7	50.9	481	2009
197	**Coconut cake**	Recipe	15.8	1.14	6.7	23.8	51.2	434	1815
198	**Crispie cakes**	Chocolate-coated; recipe	1.6	0.90	5.6	18.6	73.1	464	1951
199	**Fancy iced cakes,** individual	10 different types	12.7	0.66	3.8	14.9	68.8	407	1717
200	**Fruit cake,** plain *retail*	10 cakes, 4 brands	19.5	0.89	5.1	12.9	57.9	354	1490
201	rich	Recipe	17.6	0.63	3.8	11.0	59.6	341	1438
202	rich *retail*	e.g. Dundee	20.6	0.86	4.9	12.5	50.7	322	1357
203	rich, iced	Coated with marzipan and Royal icing; recipe	15.7	0.71	4.1	11.4	62.7	356	1504
204	wholemeal	Recipe	21.5	1.01	6.0	15.7	52.8	363	1525
205	**Gateau**	Recipe. Ref. 7	35.1	0.93	5.7	16.8	43.4	337	1413
206	**Gingerbread**	Recipe	17.6	0.97	5.7	12.6	64.7	379	1597

Cakes

Carbohydrate fractions, g per 100g

No. 11-	Food	Starch	Total sugars	Individual sugars					Dietary fibre		Fibre fractions				
				Gluc	Fruct	Sucr	Malt	Lact	Southgate method	Englyst method	Cellulose	Non-cellulosic polysaccharide Soluble	Non-cellulosic polysaccharide Insoluble	Lignin	Resistant starch
189	**All-Bran loaf**	14.8	43.6	10.9	10.8	20.2	0.2	1.3	5.4	4.6	0.9	1.1	2.6	N	N
190	**Battenburg cake**	16.1	34.0	0.7	0.5	31.8	0.1	0.9	1.5	N	N	0.4	0.6	N	N
191	**Cake mix**	35.6	41.6	0.1	Tr	39.1	2.3	0	4.9	N	N	N	N	0.1	N
192	*made up*	24.2	28.3	Tr	Tr	26.6	1.6	0	3.3	N	N	N	N	0.1	N
193	**Cherry cake**	21.8	40.0	8.0	4.4	20.7	6.9	0	0.9	1.1	N	N	N	N	N
194	**Chinese cakes and biscuits**	25.7	26.2	N	N	N	N	N	N	N	N	N	N	N	N
195	**Chocolate cake**	21.7	28.7	0.2	0.2	28.4	Tr	0	3.8	N	N	N	N	N	N
196	*with butter icing*	16.6	34.3	0.1	0.1	34.0	Tr	0	2.9	N	N	N	N	N	N
197	**Coconut cake**	29.6	21.6	0.5	0.5	20.0	0.1	0.5	3.4	2.5	0.2	0.7	1.6	N	N
198	**Crispie cakes**	32.4	40.7	0.4	0.4	39.3	0.1	0.5	1.6	0.3	0.1	0.1	0.1	N	N
199	**Fancy iced cakes,** individual	14.8	54.0	4.2	2.0	47.8	0	0	2.2	N	N	N	N	N	N
200	**Fruit cake, plain** *retail*	14.8	43.1	11.3	11.3	20.5	0	0	2.5	N	N	N	N	N	N
201	**rich**	11.2	48.4	15.8	13.0	14.9	4.7	0	3.2	1.7	0.6	0.7	0.4	(0.1)	N
202	**rich** *retail*	11.9	38.8	12.6	12.4	11.6	1.8	0.3	3.4	(1.7)	(0.6)	(0.7)	(0.4)	1.0	N
203	**rich, iced**	7.5	55.1	11.5	9.3	30.8	3.4	0	3.1	1.7	0.5	0.5	0.6	(0.1)	N
204	**wholemeal**	23.5	29.3	5.3	4.6	17.9	1.1	0.4	3.0	2.4	0.5	0.7	1.2	N	N
205	**Gateau**	11.0	32.3	0.1	0.1	31.0	Tr	1.1	0.5	0.4	Tr	0.2	0.2	Tr	N
206	**Gingerbread**	29.8	34.9	0.2	0.2	33.9	0.1	0.5	1.4	1.2	0.1	0.6	0.6	N	N

Cakes

Inorganic constituents per 100g

No. 11-	Food	Na	K	Ca	Mg	P	Fe	Cu	Zn	S	Cl	Mn	Se	I
						mg						mg	µg	
189	**All-Bran loaf**	230	470	85	69	110	2.5	0.35	1.5	N	380	N	N	N
190	**Battenburg cake**	440	140	87	24	190	1.1	0.09	0.7	N	500	0.22	4	21
191	**Cake mix**	510	89	73	10	340	0.8	0.18	0.4	96	130	0.27	N	12
192	*made up*	370	82	59	9	260	0.9	0.14	0.5	95	110	0.19	N	17
193	**Cherry cake**	310	74	72	10	110	1.3	0.10	0.5	N	400	0.16	N	1210
194	**Chinese cakes and biscuits**	170	110	86	36	130	1.5	0.15	0.7	N	130	0.31	N	N
195	**Chocolate cake**	430	190	75	46	210	1.9	0.38	1.2	N	420	0.20	5	34
196	*with butter icing*	420	140	58	36	160	1.5	0.30	0.9	N	460	0.15	4	29
197	**Coconut cake**	460	170	99	20	250	1.5	0.17	0.7	N	420	0.49	N	20
198	**Crispie cakes**	450	230	30	75	120	4.0	0.47	0.4	N	870	0.50	N	N
199	**Fancy iced cakes**, individual	250	170	44	30	120	1.4	0.25	0.7	N	230	N	N	N
200	**Fruit cake**, plain *retail*	250	390	60	25	110	1.7	0.25	0.5	N	320	N	N	150
201	*rich*	200	380	82	21	71	1.9	0.25	0.5	(67)	240	0.43	2	N
202	*rich retail*	220	410	84	29	120	3.2	0.22	0.6	67	250	0.32	N	N
203	*rich, iced*	140	330	75	33	84	1.6	0.19	0.6	N	170	0.41	2	N
204	*wholemeal*	310	240	85	33	220	1.9	0.22	1.0	N	250	0.94	11	N
205	**Gateau**	56	88	60	8	94	0.9	0.05	0.6	N	79	0.09	N	17
206	**Gingerbread**	200	160	81	14	85	1.6	0.10	0.6	N	230	0.49	N	N

No. 11-	Food	Retinol equiv µg	Vitamin D µg	Vitamin E mg	Thiamin mg	Ribo-flavin mg	Niacin mg	Trypt 60 mg	Vitamin B6 mg	Vitamin B12 µg	Folate µg	Panto-thenate µg	Biotin µg	Vitamin C mg
189	**All-Bran loaf**	16	0.41	0.46	0.18	0.24	2.6	1.0	0.28	Tr	22	0.1	2	0
190	**Battenburg cake**	46	0.34	2.68	0.08	0.16	0.5	1.4	0.05	1	13	0.4	5	0
191	**Cake mix**	N	N	N	0.26	0.01	0.7	1.0	0.04	Tr	12	N	N	0
192	*made up*	N	N	N	0.14	0.07	0.5	1.3	0.03	Tr	8	N	N	0
193	**Cherry cake**	170	1.56	1.72	0.07	0.09	0.4	1.2	0.04	1	8	0.3	5	0
194	**Chinese cakes and biscuits**	N	N	N	0.08	0.12	0.6	1.4	N	Tr	N	N	N	N
195	**Chocolate cake**	270	2.46	2.72	0.09	0.12	0.6	1.9	0.06	1	11	0.5	8	0
196	*with butter icing*	300	2.81	3.02	0.07	0.09	0.4	1.5	0.05	1	9	0.4	6	0
197	**Coconut cake**	190	1.71	2.05	0.11	0.10	0.7	1.6	0.16	Tr	11	0.4	5	0
198	**Crispie cakes**	4	1.04	0.75	0.41	0.61	6.2	0.9	0.68	1	99	0.4	5	0
199	**Fancy iced cakes**, *individual*	N	0	N	0.01	0.04	0.2	0.8	N	0	N	N	N	0
200	**Fruit cake**, *plain retail*	N	N	N	0.08	0.07	0.6	1.0	N	0	N	N	N	0
201	*rich*	120	1.07	1.26	0.08	0.07	0.6	0.8	0.11	Tr	8	0.2	5	0
202	*rich retail*	N	N	(1.30)	0.07	0.09	0.5	1.0	0.07	Tr	8	(0.2)	(5)	0
203	*rich, iced*	81	0.73	2.34	0.07	0.13	0.5	0.8	0.08	Tr	13	0.2	4	0
204	*wholemeal*	160	1.47	1.85	0.12	0.09	1.3	1.3	0.12	Tr	11	0.3	5	0
205	**Gateau**	260	0.42	0.77	0.07	0.18	0.3	1.5	0.05	1	11	0.5	8	0
206	**Gingerbread**	130	1.18	1.36	0.10	0.08	0.6	1.3	0.06	Tr	8	0.3	4	0

No. 11-	Food	Description and main data sources	Water g	Total nitrogen g	Protein g	Fat g	Carbohydrate g	Energy value kcal	kJ
207	**Glutinous rice flour cakes**, Chinese	3 assorted samples	29.8	0.61	3.5	6.2	58.8	290	1230
208	**Lardy cake**	Recipe	23.7	1.16	6.7	16.3	53.7	375	1577
209	**Madeira cake**	10 cakes, 4 brands	20.2	0.94	5.4	16.9	58.4	393	1652
210	**Rock cakes**	Recipe	15.2	0.91	5.4	16.4	60.5	396	1665
211	**Sponge cake**	Basic recipe, creaming method	15.2	1.06	6.4	26.3	52.4	459	1920
212	*fatless*	Basic recipe, whisking method	31.5	1.65	10.1	6.1	53.0	294	1245
213	*jam filled*	10 cakes, 3 brands; sandwich and Swiss roll	24.5	0.74	4.2	4.9	64.2	302	1280
214	*with butter icing*	Recipe. Ref. 7	13.0	0.74	4.5	30.6	52.4	490	2046
215	*frozen*	Cream filled, 10 samples, 2 brands	35.4	0.60	3.4	16.6	40.8	316	1325
216	**Swiss roll**	Recipe	32.9	1.17	7.2	4.4	55.5	276	1172
217	**Swiss rolls**, chocolate, individual	10 samples, 5 brands, 4 bakeries	17.5	0.75	4.3	11.3	58.1	337	1421
218	**Welsh cakes**	Recipe	12.0	0.94	5.6	19.6	61.8	431	1808

No. 11-	Food	Starch	Total sugars	Individual sugars					Dietary fibre		Fibre fractions				
									Southgate method	Englyst method		Non-cellulosic polysaccharide			Resistant starch
				Gluc	Fruct	Sucr	Malt	Lact			Cellulose	Soluble	Insoluble	Lignin	
207	**Glutinous rice flour cakes,**														
	Chinese	26.8	32.0	N	N	N	N	N	N	N	N	N	N	N	N
208	**Lardy cake**	43.7	10.0	0.3	0.3	7.6	0.1	1.7	2.2	1.8	0.1	0.9	0.9	N	N
209	**Madeira cake**	21.9	36.5	0.5	0.5	35.5	0	0	1.3	0.9	N	N	N	Tr	0.2
210	**Rock cakes**	28.3	32.2	6.4	6.2	19.1	0.1	0.4	1.7	1.5	0.2	0.7	0.6	N	N
211	**Sponge cake**	22.0	30.4	0.2	0.2	30.0	0.1	0	1.0	0.9	Tr	0.4	0.4	Tr	N
212	*fatless*	22.1	30.9	0.1	0.1	30.5	0.1	0	1.0	0.9	Tr	0.4	0.4	Tr	N
213	*jam filled*	16.5	47.7	8.1	3.9	35.7	0	0	1.1	1.8	0.1	N	N	Tr	Tr
214	*with butter icing*	15.4	37.1	0.1	0.1	36.7	0.1	0	0.7	0.6	Tr	0.3	0.3	Tr	N
215	*frozen*	13.2	27.6	4.3	1.2	18.9	1.9	1.2	1.3	N	Tr	N	N	0.1	N
216	**Swiss roll**	14.5	41.0	6.7	3.7	28.6	2.0	0	1.0	0.8	Tr	0.5	0.3	Tr	N
217	**Swiss rolls,** chocolate, individual	16.3	41.8	4.8	0.2	32.5	2.4	1.9	2.4	N	N	N	N	0.6	N
218	**Welsh cakes**	30.2	31.6	5.2	5.1	20.7	0.1	0.5	1.7	1.5	0.2	0.7	0.6	N	N

No. 11-	Food	Na	K	Ca	Mg	P	Fe	Cu	Zn	S	Cl	Mn	Se	I
						mg							µg	
207	**Glutinous rice flour cakes,**													
	Chinese	8	62	12	9	44	1.9	0.13	0.7	N	49	0.43	N	N
208	**Lardy cake**	270	150	130	19	100	1.4	0.12	0.6	N	460	0.34	2	N
209	**Madeira cake**	380	120	42	12	120	1.1	0.10	0.5	N	500	N	N	N
210	**Rock cakes**	390	210	110	15	240	1.2	0.26	0.6	N	270	0.35	3	15
211	**Sponge cake**	350	82	66	9	150	1.2	0.10	0.7	N	410	0.18	5	25
212	*fatless*	82	120	75	13	150	1.7	0.10	1.0	N	120	0.19	10	34
213	*jam filled*	420	140	44	14	220	1.6	0.20	0.5	N	260	N	(10)	14
214	*with butter icing*	360	59	47	7	110	0.9	0.08	0.6	N	470	0.13	4	21
215	*frozen*	300	120	44	11	180	0.7	0.14	0.4	100	580	0.14	N	N
216	**Swiss roll**	130	110	98	11	180	1.5	0.11	0.7	N	86	N	7	26
217	**Swiss rolls,** chocolate,													
	individual	350	210	77	19	200	1.1	0.25	0.5	120	510	0.23	N	13
218	**Welsh cakes**	200	190	96	15	130	1.2	0.21	0.5	N	240	0.34	3	N

Cakes *continued*

No. 11-	Food	Retinol equiv µg	Vitamin D µg	Vitamin E mg	Thiamin mg	Ribo-flavin mg	Niacin mg	Trypt 60 mg	Vitamin B6 mg	Vitamin B12 µg	Folate µg	Panto-thenate µg	Biotin µg	Vitamin C mg
207	Glutinous rice flour cakes, Chinese	N	N	N	Tr	0.02	0.9	0.8	N	Tr	N	N	N	N
208	Lardy cake	20	0.01	0.21	0.16	0.09	1.1	1.4	0.09	Tr	18	0.3	2	0
209	Madeira cake	N	N	N	0.06	0.11	0.5	1.1	N	Tr	N	N	N	0
210	Rock cakes	170	1.55	1.71	0.12	0.07	0.7	1.1	0.08	Tr	7	0.2	4	0
211	Sponge cake	280	2.61	2.83	0.09	0.12	0.5	1.6	0.06	1	10	0.5	7	0
212	fatless	110	0.70	1.02	0.11	0.24	0.5	2.7	0.08	1	18	0.9	15	0
213	jam filled	N	N	Tr	0.04	0.07	0.4	0.9	N	(1)	N	N	N	0
214	with butter icing	320	3.02	3.18	0.06	0.08	0.3	1.1	0.04	1	7	0.3	5	0
215	frozen	(260)	(0.40)	Tr	0.09	0.10	0.3	0.7	0.05	Tr	6	N	N	0
216	Swiss roll	79	0.50	0.73	0.07	0.17	0.3	1.9	0.05	1	13	0.6	11	1
217	Swiss rolls, chocolate individual	N	N	N	0.12	0.19	0.3	0.9	0.03	Tr	10	N	N	0
218	Welsh cakes	110	0.18	0.46	0.12	0.07	0.8	1.2	0.08	Tr	7	0.2	4	Tr

No. 11-	Food	Description and main data sources	Water g	Total nitrogen g	Protein g	Fat g	Carbohydrate g	Energy value kcal	kJ
219	**Cheese pastry** *cooked*	Recipe. Ref. 7	14.5	2.15	13.2	33.9	37.2	498	2073
220	**Choux pastry** *raw*	Recipe	61.6	0.92	5.5	12.9	19.4	211	881
221	*cooked*	Recipe	40.9	1.41	8.5	19.8	29.8	325	1355
222	**Flaky pastry** *raw*	Recipe	30.1	0.74	4.2	30.7	34.8	424	1765
223	*cooked*	Recipe	7.7	0.97	5.6	40.6	45.9	560	2332
224	**Puff pastry** *frozen, raw*	10 samples, 2 brands	31.8	1.00	5.7	23.5	37.0	373	1558
225	**Shortcrust pastry** *raw*	Recipe	20.0	0.99	5.7	27.9	46.8	449	1874
226	*cooked*	Recipe	7.2	1.15	6.6	32.3	54.2	521	2174
227	*frozen, raw*	10 samples, 2 brands	20.9	0.79	4.5	28.4	44.3	440	1836
228	**Wholemeal pastry** *raw*	Recipe. Ref. 7	20.0	1.32	7.7	28.4	38.5	431	1797
229	*cooked*	Recipe. Ref. 7	7.4	1.52	8.9	32.9	44.6	499	2080

Carbohydrate fractions, g per 100g

| No. 11- | Food | Starch | Total sugars | Individual sugars | | | | | Dietary fibre | | Fibre fractions | | | | Resistant starch |
| | | | | Gluc | Fruct | Sucr | Malt | Lact | Southgate method | Englyst method | | Non-cellulosic polysaccharide | | | |
											Cellulose	Soluble	Insoluble	Lignin	
219	**Cheese pastry** *cooked*	36.5	0.8	0.3	0.3	0.1	0.1	Tr	1.7	1.5	(0.1)	0.7	0.7	Tr	N
220	**Choux pastry** *raw*	19.0	0.4	0.1	0.1	0.1	0.1	0	0.9	0.8	Tr	0.4	0.4	Tr	N
221	*cooked*	29.2	0.6	0.2	0.2	0.1	0.1	0	1.4	1.2	Tr	0.6	0.6	Tr	N
222	**Flaky pastry** *raw*	34.1	0.7	0.2	0.3	0.1	0.1	0	1.6	1.4	0.1	0.7	0.7	Tr	N
223	*cooked*	45.0	0.9	0.3	0.3	0.2	0.1	0	2.1	1.8	0.1	0.9	0.9	Tr	N
224	**Puff pastry** *frozen, raw*	35.7	1.3	0.1	Tr	0.1	1.0	Tr	2.3	N	(0.1)	(0.7)	(0.7)	Tr	N
225	**Shortcrust pastry** *raw*	45.8	0.9	0.3	0.3	0.2	0.1	0	2.2	1.9	0.1	0.9	0.9	Tr	N
226	*cooked*	53.2	1.1	0.4	0.4	0.2	0.1	0	2.5	2.2	0.1	1.0	1.0	Tr	(0.2)
227	*frozen, raw*	43.6	0.7	0.1	Tr	Tr	0.5	0	2.3	(1.9)	(0.1)	(0.9)	(0.9)	Tr	N
228	**Wholemeal pastry** *raw*	37.2	1.3	0.3	0.3	0.6	0	0	5.2	5.4	0.8	1.2	3.4	0.1	N
229	*cooked*	43.1	1.5	0.4	0.4	0.7	0	0	6.0	6.3	1.0	1.4	3.9	0.2	N

No. 11-	Food	Na	K	Ca	Mg	P	Fe	Cu	Zn	S	Cl	Mn	Se	I
						mg							µg	
219	**Cheese pastry** *cooked*	560	98	290	20	230	1.1	0.11	1.2	N	890	0.30	6	23
220	**Choux pastry** *raw*	230	70	50	9	79	1.0	0.06	0.5	N	360	0.16	5	19
221	*cooked*	360	110	77	14	120	1.6	0.10	0.8	N	550	0.24	7	29
222	**Flaky pastry** *raw*	350	71	64	11	52	1.0	0.08	0.3	N	570	0.27	2	10
223	*cooked*	460	94	84	15	69	1.3	0.11	0.5	N	750	0.36	2	13
224	**Puff pastry** *frozen, raw*	310	70	58	12	54	0.9	0.14	0.4	95	540	0.24	N	N
225	**Shortcrust pastry** *raw*	410	91	85	15	68	1.3	0.10	0.4	N	680	0.36	2	11
226	*cooked*	480	110	99	17	79	1.5	0.12	0.5	N	790	0.42	3	13
227	*frozen, raw*	160	97	43	12	58	0.8	0.20	0.5	96	290	0.29	N	N
228	**Wholemeal pastry** *raw*	360	210	24	74	190	2.4	0.28	1.8	N	570	1.89	32	N
229	*cooked*	410	240	28	86	230	2.8	0.32	2.1	N	650	2.19	37	N

No. 11-	Food	Retinol equiv μg	Vitamin D μg	Vitamin E mg	Thiamin mg	Ribo-flavin mg	Niacin mg	Trypt 60 mg	Vitamin B6 mg	Vitamin B12 μg	Folate μg	Panto-thenate μg	Biotin μg	Vitamin C mg
219	**Cheese pastry** *cooked*	210	1.04	1.27	0.12	0.14	0.8	2.9	0.08	Tr	10	0.2	1	Tr
220	**Choux pastry** *raw*	150	1.29	1.47	0.10	0.12	0.4	1.4	0.06	1	18	0.5	7	0
221	*cooked*	230	1.98	2.26	0.12	0.16	0.7	2.2	0.07	1	14	0.6	10	0
222	**Flaky pastry** *raw*	130	1.33	1.47	0.14	0.01	0.8	0.9	0.07	0	10	0.1	1	1
223	*cooked*	180	1.76	1.95	0.14	0.02	1.0	1.1	0.07	0	7	0.1	1	Tr
224	**Puff pastry** *frozen, raw*	0	N	N	0.19	0.04	0.9	1.2	0.05	Tr	16	N	N	0
225	**Shortcrust pastry** *raw*	120	1.19	1.38	0.19	0.02	1.0	1.1	0.09	0	13	0.2	1	0
226	*cooked*	140	1.39	1.60	0.16	0.02	1.1	1.3	0.08	0	8	0.2	1	0
227	*frozen, raw*	0	0	N	0.17	Tr	0.8	0.9	0.06	Tr	16	0.2	1	0
228	**Wholemeal pastry** *raw*	120	1.20	2.05	0.28	0.05	3.4	1.5	0.30	0	34	0.5	4	0
229	*cooked*	140	1.38	2.37	0.25	0.05	3.8	1.7	0.26	0	20	0.4	5	0

Buns and pastries

Composition of food per 100g

No. 11-	Food	Description and main data sources	Water g	Total nitrogen g	Protein g	Fat g	Carbohydrate g	Energy value kcal	kJ
230	**Asian pastries**	5 samples, assorted sata pastries	6.9	1.11	6.3	39.1	43.0	538	2242
231	**Burfi**	5 assorted samples, fudge-type sweet	16.3	1.86	11.6	19.9	17.7	292	1217
232	**Chelsea buns/Bath buns**	Recipe. Ref. 7	20.1	1.34	7.8	13.8	56.1	366	1542
233	**Choux buns**	Filled with fresh cream. Recipe	43.9	0.90	5.4	32.5	17.6	381	1577
234	**Cream horns**	Recipe. Ref. 7	34.4	0.66	3.8	35.8	25.8	435	1803
235	**Crumpets** *fresh*	10 samples, 4 brands, 4 bakeries	52.4	1.05	6.0	0.9	38.6	177	753
236	*toasted*	Calculated using 11% weight loss	46.5	1.18	6.7	1.0	43.4	199	846
237	**Currant buns**	10 samples, 5 brands, 5 bakeries	27.7	1.34	7.6	7.5	52.7	296	1250
238	**Custard tart** *large*	Recipe	48.9	0.94	5.7	16.7	28.5	280	1171
239	**Custard tarts**, *individual*	10 samples, 2 brands, 8 bakeries	44.7	1.00	6.3	14.5	32.4	277	1161
240	**Danish pastries**	10 samples, different shops	21.6	1.01	5.8	17.6	51.3	374	1571
241	**Doughnuts**, custard-filled	Filled with confectioners' custard. Recipe	30.5	1.06	6.2	19.0	43.3	358	1500
242	jam	10 samples, different shops	26.9	1.00	5.7	14.5	48.8	336	1414
243	ring	10 samples, different shops	23.8	1.07	6.1	21.7	47.2	397	1662
244	ring, iced	Recipe	22.5	0.84	4.8	17.5	55.1	383	1610
245	**Eccles cake**	Recipe. Ref. 7	4.2	0.68	3.9	26.4	59.3	475	1991
246	**Eclairs** *fresh*	With chocolate icing and fresh cream filling; recipe	34.6	0.68	4.1	23.8	37.9	373	1559
247	*frozen*	10 samples of the same brand (Birds Eye)	38.7	0.98	5.6	30.6	26.1	396	1647

Buns and pastries

Carbohydrate fractions, g per 100g

No. 11-	Food	Starch	Total sugars	Individual sugars					Dietary fibre		Fibre fractions				
				Gluc	Fruct	Sucr	Malt	Lact	Southgate method	Englyst method	Cellulose	Non-cellulosic polysaccharide Soluble	Insoluble	Lignin	Resistant starch
230	Asian pastries	20.4	22.6	N	N	N	N	N	2.1	N	N	N	N	N	N
231	Burfi	16.5	1.2	N	N	N	N	N	N	N	N	N	N	N	N
232	Chelsea buns/Bath buns	34.7	21.4	4.7	4.5	11.2	Tr	1.0	2.2	1.7	0.2	0.8	0.7	N	N
233	Choux buns	16.1	1.5	0.1	0.1	0.1	0.1	1.2	0.8	0.7	0.1	0.3	0.3	Tr	N
234	Cream horns	19.6	6.2	2.0	1.1	1.2	0.6	1.2	1.0	0.9	0.1	0.4	0.4	Tr	N
235	Crumpets *fresh*	36.9	1.7	0.1	0.2	0.1	1.3	0	2.6	(1.8)	(0.1)	(0.9)	(0.9)	Tr	(0.7)
236	*toasted*	41.5	1.9	0.1	0.2	0.1	1.5	0	2.9	(2.0)	(0.1)	(1.0)	(1.0)	Tr	N
237	Currant buns	37.6	15.1	6.9	7.1	0.4	0.3	0.4	1.8	N	N	N	N	0.1	N
238	Custard tart *large*	22.6	5.9	0.1	0.1	3.2	0.1	2.3	1.1	0.9	Tr	0.4	0.4	N	N
239	Custard tarts, individual	19.6	12.8	0.1	0.1	10.5	0.5	1.6	1.2	(1.2)	Tr	N	N	0.1	Tr
240	Danish pastries	22.8	28.5	6.3	5.5	13.9	2.2	0.6	2.7	1.6	0.2	N	N	0.3	0.3
241	Doughnuts, custard-filled	27.4	15.9	1.9	2.3	10.1	0.9	0.7	2.6	N	N	N	N	0.2	N
242	jam	30.0	18.8	5.3	4.2	6.4	2.6	0.2	2.5	N	N	N	N	0.3	N
243	ring	31.9	15.3	2.3	2.7	9.1	1.1	0.1	3.1	N	N	N	N	0.2	N
244	ring, iced	24.4	30.7	1.8	2.1	25.9	0.8	0.1	2.4	N	N	N	N	0.1	N
245	Eccles cake	18.2	41.1	15.9	15.4	9.8	Tr	0	2.0	1.6	0.4	0.8	0.5	N	N
246	Eclairs *fresh*	11.5	26.4	0.1	0.1	25.4	Tr	0.8	0.5	0.5	(0.1)	0.2	0.2	(0.2)	N
247	*frozen*	19.5	6.6	Tr	Tr	5.3	0.3	0.9	1.5	0.8	0.1	(0.3)	0.2	0.2	0.3

Buns and pastries

Inorganic constituents per 100g

No. 11-	Food	Na	K	Ca	Mg	P	Fe	Cu	Zn	S	Cl	Mn	Se	I
		mg											µg	
230	**Asian pastries**	190	170	160	19	130	0.9	0.15	0.6	N	50	0.20	N	N
231	**Burfi**	210	490	340	42	300	0.9	0.10	1.3	N	330	0.08	N	N
232	**Chelsea buns/Bath buns**	330	220	110	25	120	1.5	0.26	0.8	78	530	0.41	N	N
233	**Choux buns**	210	88	65	10	89	1.0	0.05	0.5	N	330	0.13	N	16
234	**Cream horns**	200	71	62	11	57	0.7	0.07	0.4	N	310	0.18	N	6
235	**Crumpets** *fresh*	720	83	110	16	160	1.0	0.18	0.5	95	870	0.30	(24)	(1)
236	*toasted*	810	93	120	18	180	1.1	0.20	0.6	110	980	0.30	(27)	(1)
237	**Currant buns**	230	210	110	27	100	1.9	0.18	0.6	110	210	0.41	N	N
238	**Custard tart** *large*	250	130	110	14	100	0.9	0.06	0.5	N	400	0.18	N	22
239	**Custard tarts**, individual	130	110	95	14	98	0.8	0.07	0.5	N	390	0.16	N	N
240	**Danish pastries**	190	170	92	24	98	1.3	0.06	0.5	74	340	0.35	N	N
241	**Doughnuts**, custard-filled	200	99	83	20	90	1.1	0.12	0.6	72	320	0.28	N	(19)
242	jam	180	110	72	19	71	1.2	0.09	0.5	78	290	0.31	N	15
243	ring	230	87	76	21	81	1.2	0.14	0.6	73	360	0.33	N	(17)
244	ring, iced	170	76	59	19	66	1.0	0.13	0.5	55	280	0.27	N	N
245	**Eccles cake**	240	360	79	22	66	1.2	0.43	0.5	45	380	N	N	N
246	**Eclairs** *fresh*	150	88	47	16	73	0.9	0.11	0.4	N	250	0.14	N	24
247	*frozen*	73	160	87	20	120	1.1	0.22	0.8	82	75	0.15	N	N

Buns and pastries

No. 11-	Food	Retinol equiv µg	Vitamin D µg	Vitamin E mg	Thiamin mg	Riboflavin mg	Niacin mg	Trypt 60 mg	Vitamin B6 mg	Vitamin B12 µg	Folate µg	Pantothenate µg	Biotin µg	Vitamin C mg
230	Asian pastries	N	N	N	0.09	0.16	0.4	1.2	N	0	N	N	N	0
231	Burfi	200	0.30	0.30	0.13	0.80	0.2	2.8	N	1	4	N	N	0
232	Chelsea buns/Bath buns	17	0.10	1.49	0.16	0.13	1.4	1.6	0.11	Tr	32	0.4	6	Tr
233	Choux buns	330	0.38	1.74	0.07	0.16	0.4	1.4	0.05	1	11	0.4	6	Tr
234	Cream horns	210	0.07	2.17	0.07	0.06	0.5	0.8	0.04	Tr	7	0.1	1	1
235	Crumpets *fresh*	0	0	(0.17)	0.18	0.03	0.9	1.2	0.05	Tr	8	(0.3)	(4)	0
236	*toasted*	0	0	(0.19)	0.17	0.03	1.0	1.4	0.06	Tr	9	(0.3)	(4)	0
237	Currant buns	N	0	N	0.37	0.16	1.5	1.6	0.11	Tr	40	N	N	0
238	Custard tart *large*	110	0.72	0.89	0.10	0.13	0.5	1.3	0.07	Tr	10	0.4	4	Tr
239	Custard tarts, individual	(32)	N	N	0.14	0.16	0.5	1.4	0.03	Tr	13	N	N	0
240	Danish pastries	N	N	Tr	0.13	0.07	0.9	1.2	0.07	Tr	20	(0.5)	(7)	0
241	Doughnuts, custard-filled	N	N	0.09	0.20	0.10	1.0	2.0	0.03	Tr	18	N	N	N
242	jam	N	N	Tr	0.22	0.07	1.3	1.2	0.03	Tr	21	N	N	N
243	ring	N	N	Tr	0.22	0.07	1.2	1.2	0.02	Tr	19	N	N	0
244	Doughnuts, ring, iced	N	N	0.03	0.17	0.06	0.9	1.6	0.02	Tr	15	N	N	0
245	Eccles cake	61	0.05	1.99	0.11	0.03	0.9	0.7	0.11	0	5	0.1	2	0
246	Eclairs *fresh*	210	0.25	1.25	0.06	0.11	0.3	1.0	0.03	Tr	8	0.3	5	Tr
247	*frozen*	240	(0.25)	(1.25)	0.10	0.19	0.3	1.1	0.03	1	11	(0.3)	(5)	Tr

Composition of food per 100g

No. 11-	Food	Description and main data sources	Water g	Total nitrogen g	Protein g	Fat g	Carbohydrate g	Energy value kcal	kJ
248	**Greek pastries**	4 assorted samples, baclava, tangos, tsamika, shredded type	17.5	0.82	4.7	17.0	40.0	322	1349
249	**Gulab jamen/gulab jambu**								
	homemade	Recipe	26.4	0.96	5.8	11.7	56.7	341	1439
250	*retail*	Recipe	25.4	1.23	7.6	15.0	51.4	358	1506
251	**Halva**	Greek sweet, 10 assorted samples	2.4	2.37	14.8	31.9	54.2	615	2569
252	**Halwa**	Asian sweet, 5 samples, assorted colours and flavours	16.4	0.29	1.8	13.2	68.0	381	1607
253	**Hot cross buns**	Recipe	25.2	1.27	7.4	6.8	58.5	310	1313
254	**Jam tarts**	Recipe	19.6	0.57	3.3	14.9	62.0	380	1598
255	*retail*	10 samples, 6 brands, 4 bakeries	14.4	0.53	3.3	13.0	63.4	368	1551
256	*wholemeal*	Recipe. Ref. 7	19.6	0.74	4.4	15.1	57.5	369	1554
257	**Jellabi**	Recipe	24.9	0.68	4.0	13.5	60.3	363	1528
258	**Mince pies**, individual	Recipe	12.0	0.75	4.3	20.4	59.0	423	1772
259	**Mincemeat tart** *one crust*	Recipe. Ref. 7	17.5	0.59	3.4	15.8	57.8	373	1566
260	**Muffins**	Recipe	33.9	1.73	10.1	6.3	49.6	283	1199
261	*bran*	Recipe	30.1	1.29	7.8	7.7	45.6	272	1148
262	**Pinni/dabra**	Asian sweetmeat. Recipe. Ref. 6	4.1	0.91	5.0	31.6	58.2	523	2186
263	**Rum baba**	Recipe	53.2	0.62	3.5	8.1	32.2	223	939

Buns and pastries *continued*

Carbohydrate fractions, g per 100g

No. 11-	Food	Starch	Total sugars	Individual sugars					Dietary fibre		Fibre fractions				Resistant starch
				Gluc	Fruct	Sucr	Malt	Lact	Southgate method	Englyst method	Cellulose	Non-cellulosic polysaccharide Soluble	Insoluble	Lignin	
248	**Greek pastries**	21.6	18.4	N	N	N	N	N	1.9	N	N	N	N	N	N
249	**Gulab jamen/gulab jambu**														
250	*homemade*	16.2	40.5	Tr	Tr	33.9	0	6.5	0.8	0.6	Tr	0.3	0.3	N	N
	retail	5.1	46.3	Tr	Tr	34.9	0	11.3	0.4	0.2	Tr	Tr	0.1	N	N
251	**Halva**	1.1	53.1	N	N	N	N	N	N	N	N	N	N	N	N
252	**Halwa**	9.0	59.0	N	N	N	N	N	N	N	N	N	N	N	N
253	**Hot cross buns**	35.1	23.4	5.2	4.4	11.7	1.2	0.9	2.2	1.7	0.1	0.8	0.7	N	N
254	**Jam tarts**	24.5	37.5	15.3	8.4	9.1	4.7	0	1.7	1.6	0.1	0.9	0.5	0.1	N
255	*retail*	27.4	36.0	12.4	5.9	11.0	6.6	0.1	2.5	N	N	N	N	0.1	N
256	*wholemeal*	19.8	37.7	15.3	8.4	9.4	4.6	0	3.3	3.4	0.5	1.0	1.8	0.1	N
257	**Jellabi**	34.5	25.8	0.2	0.2	25.3	0.1	0	1.5	1.2	0.1	0.5	0.6	Tr	N
258	**Mince pies**, individual	30.9	28.1	14.1	13.0	0.1	0.9	0	2.8	2.1	N	N	N	N	N
259	**Mincemeat tart** *one crust*	22.5	35.3	17.8	16.3	0.1	1.1	0	2.8	2.0	N	N	N	N	N
260	**Muffins**	47.0	2.6	0.4	0.4	0.2	Tr	1.7	2.7	2.0	0.1	1.0	1.0	Tr	N
261	*bran*	26.1	19.5	0.2	0.1	17.3	0.1	1.8	8.5	7.7	1.4	1.0	5.3	0.6	N
262	**Pinni/dabra**	17.1	41.1	0.3	0.3	40.4	Tr	0	3.2	2.3	0.4	0.5	1.3	N	N
263	**Rum baba**	12.5	19.7	0.2	0.2	19.4	Tr	0	1.3	0.8	0.1	0.2	0.4	N	N

Buns and pastries *continued*

Inorganic constituents per 100g

No. 11-	Food	Na	K	Ca	Mg	P	Fe	Cu	Zn	S	Cl	Mn	Se	I
		mg											µg	
248	**Greek pastries**	310	90	44	22	70	0.9	0.14	0.4	N	390	0.26	N	N
249	**Gulab jamen/gulab jambu**													
250	homemade	120	200	180	16	180	0.4	0.04	0.3	N	150	0.13	N	N
	retail	130	320	230	25	210	0.2	0.02	0.3	77	240	0.06	N	N
251	**Halva**	190	190	110	170	330	5.7	1.40	4.4	N	210	0.93	N	N
252	**Halwa**	27	67	30	17	49	2.1	0.19	0.3	N	110	0.13	N	N
253	**Hot cross buns**	120	200	110	24	110	1.6	0.24	0.7	71	190	0.41	N	N
254	**Jam tarts**	230	110	55	12	46	1.4	0.14	0.3	N	370	N	1	10
255	retail	130	120	72	14	50	1.7	0.18	0.6	59	160	0.35	N	2
256	wholemeal	280	170	22	44	110	2.0	0.24	1.1	N	430	N	17	N
257	**Jellabi**	2	64	52	8	48	0.8	0.07	0.4	N	32	0.29	2	4
258	**Mince pies,** individual	310	180	75	15	58	1.5	0.15	0.3	N	550	0.34	N	N
259	**Mincemeat tart** one crust	240	200	64	14	49	1.5	0.16	0.3	N	450	0.30	N	N
260	**Muffins**	130	180	140	29	150	1.8	0.20	0.9	96	230	0.43	28	N
261	bran	770	350	120	110	370	3.3	0.33	3.5	N	450	1.87	3	N
262	**Pinni/dabra**	2	160	61	48	110	1.3	0.10	0.9	N	12	0.67	N	N
263	**Rum baba**	120	94	39	18	65	0.9	0.10	0.5	N	190	0.17	2	8

Buns and pastries *continued*

No. 11-	Food	Retinol equiv μg	Vitamin D μg	Vitamin E mg	Thiamin mg	Ribo-flavin mg	Niacin mg	Trypt 60 mg	Vitamin B6 mg	Vitamin B12 μg	Folate μg	Panto-thenate μg	Biotin μg	Vitamin C mg
248	**Greek pastries**	N	N	N	0.09	0.04	1.0	1.0	N	N	N	N	N	N
249	**Gulab jamen/gulab jambu**													
250	*homemade*	59	1.51	1.95	0.05	0.12	0.2	1.3	0.05	Tr	6	0.3	2	1
250	*retail*	100	2.80	2.22	0.05	0.22	0.1	1.8	0.07	1	8	0.4	3	2
251	**Halva**	N	N	N	0.34	0.13	2.0	2.8	N	Tr	N	N	N	N
252	**Halva**	65	N	N	0.02	0.04	0.1	0.3	N	Tr	15	N	N	N
253	**Hot cross buns**	68	0.53	0.71	0.15	0.11	1.3	1.5	0.10	Tr	28	0.3	5	0
254	**Jam tarts**	64	0.64	0.74	0.07	0.01	0.5	0.6	0.04	0	4	0.1	Tr	2
255	*retail*	N	N	N	0.06	0.02	0.5	0.7	0.03	Tr	5	N	N	Tr
256	*wholemeal*	63	0.64	1.09	0.11	0.02	1.7	0.8	0.12	0	9	0.2	2	2
257	**Jellabi**	0	0	3.23	0.07	0.01	0.4	0.8	0.05	0	5	0.1	Tr	0
258	**Mince pies**, individual	81	0.80	0.93	0.11	0.02	0.8	0.8	0.08	0	6	0.1	Tr	0
259	**Mincemeat tart** *one crust*	59	0.58	0.68	0.09	0.02	0.7	0.6	0.08	0	6	0.1	Tr	0
260	**Muffins**	67	0.42	0.67	0.20	0.16	1.8	2.1	0.12	Tr	45	0.5	7	Tr
261	*bran*	81	0.16	0.86	0.21	0.16	5.8	1.8	0.25	Tr	31	0.6	12	Tr
262	**Pinni/dabra**	190	0.49	5.27	0.09	0.08	1.1	0.9	0.06	0	12	0.1	1	0
263	**Rum baba**	14	0.09	1.47	0.07	0.10	0.7	0.8	0.04	Tr	26	0.2	4	1

Buns and pastries *continued*

Composition of food per 100g

No. 11-	Food	Description and main data sources	Water g	Total nitrogen g	Protein g	Fat g	Carbohydrate g	Energy value kcal	Energy value kJ
264	**Scones**, cheese	Recipe. Ref. 7	26.9	1.67	10.1	17.8	43.2	363	1522
265	fruit	10 samples, 2 brands, 8 bakeries	25.3	1.28	7.3	9.8	52.9	316	1333
266	plain	Recipe	22.9	1.23	7.2	14.6	53.8	362	1523
267	potato	Recipe	39.5	0.88	5.1	14.3	39.1	296	1241
268	wholemeal	Recipe. Ref. 7	26.9	1.48	8.7	14.4	43.1	326	1368
269	wholemeal, fruit	Recipe	24.1	1.37	8.1	12.8	47.2	324	1365
270	**Scotch pancakes**	Drop scones; recipe	39.1	0.99	5.8	11.7	43.6	292	1228
271	**Sevyiaan**	Sweet Indian snack, recipe	21.8	0.78	4.8	27.6	44.4	442	1837
272	**Strawberry tartlets**	Recipe	59.4	0.43	2.5	10.8	26.2	206	863
273	**Teacakes** *fresh*	10 samples, 4 brands, 6 bakeries	26.7	1.40	8.0	7.5	52.5	296	1253
274	*toasted*	Calculated using weight loss of 10%	18.6	1.56	8.9	8.3	58.3	329	1392
275	**Vanilla slices**	Recipe	37.6	0.74	4.5	17.9	40.2	330	1383
276	**Waffles**	Recipe	33.1	1.45	8.7	16.7	39.6	334	1401
277	**Welsh cheesecakes**	Recipe	19.2	0.90	5.3	23.4	52.0	427	1788

Buns and pastries *continued*

Carbohydrate fractions, g per 100g

No. 11-	Food	Starch	Total sugars	Gluc	Fruct	Sucr	Malt	Lact	Southgate method	Englyst method	Cellulose	Soluble	Insoluble	Lignin	Resistant starch
						Individual sugars			Dietary fibre			Non-cellulosic polysaccharide		Fibre fractions	
264	**Scones,** cheese	40.9	2.3	0.3	0.3	0.1	0.1	1.6	1.9	1.6	0.1	0.8	0.8	Tr	N
265	fruit	36.0	16.9	3.9	3.0	8.2	0.4	1.3	3.6	N	N	N	N	0.2	N
266	plain	47.9	5.9	0.7	0.7	3.3	0.1	1.8	2.2	1.9	0.1	0.9	0.9	Tr	N
267	potato	37.3	1.8	0.3	0.3	0.2	0.1	0.9	1.7	1.6	0.2	0.6	0.6	N	N
268	wholemeal	37.1	5.9	0.3	0.3	3.6	0	1.7	5.0	5.2	0.8	1.2	3.2	0.2	N
269	wholemeal, fruit	33.0	14.2	4.8	4.7	3.2	0	1.5	5.2	4.9	0.8	1.2	2.9	0.4	N
270	**Scotch pancakes**	34.8	8.9	0.2	0.2	6.1	0.1	2.2	1.6	1.4	0.1	0.7	0.7	Tr	N
271	**Sevyiaan**	23.7	20.8	1.3	1.2	15.7	0.6	2.0	2.2	1.2	0.2	0.5	0.5	N	N
272	**Strawberry tartlets**	18.9	7.3	1.1	1.2	5.1	Tr	0	1.5	1.2	0.2	0.5	0.4	0.3	N
273	**Teacakes** *fresh*	37.7	14.8	5.9	6.2	1.0	1.3	0.3	4.2	N	N	N	N	0.4	N
274	*toasted*	41.9	16.4	6.6	6.9	1.1	1.4	0.4	4.7	N	N	N	N	0.4	N
275	**Vanilla slices**	20.4	19.8	1.7	1.0	15.1	0.5	1.5	0.9	0.8	Tr	0.4	0.4	Tr	N
276	**Waffles**	36.0	3.7	0.3	0.2	0.1	0	3.1	2.0	1.5	0.1	0.7	0.7	Tr	N
277	**Welsh cheesecakes**	29.3	22.7	3.8	2.2	15.5	1.2	0	1.5	1.3	0.1	0.7	0.6	Tr	N

Buns and pastries *continued*

Inorganic constituents per 100g

No. 11-	Food	Na	K	Ca	Mg	P	Fe (mg)	Cu	Zn	S	Cl	Mn	Se (µg)	I
264	**Scones**, cheese	760	140	250	19	470	1.1	0.18	1.0	N	550	0.32	4	22
265	fruit	710	220	150	24	360	1.5	0.22	0.8	120	450	0.35	N	N
266	plain	770	150	180	18	460	1.3	0.19	0.7	N	450	0.36	2	19
267	potato	730	180	130	16	410	1.0	0.18	0.5	N	430	0.62	2	13
268	wholemeal	730	250	110	75	560	2.3	0.35	2.0	N	400	1.81	28	N
269	wholemeal, fruit	650	360	110	70	510	2.3	0.37	1.8	N	360	1.65	31	N
270	**Scotch pancakes**	430	300	120	16	95	1.0	0.07	0.5	N	560	0.27	2	18
271	**Sevyiaan**	250	170	70	26	90	0.9	0.12	0.8	69	450	0.36	N	N
272	**Strawberry tartlets**	160	93	39	9	35	0.7	0.07	0.2	N	270	0.25	1	N
273	**Teacakes** *fresh*	270	220	88	29	100	2.6	0.24	0.7	110	440	0.46	N	N
274	*toasted*	300	240	98	32	110	2.9	0.27	0.8	120	490	0.51	N	N
275	**Vanilla slices**	230	100	78	11	76	0.8	0.07	0.4	27	360	N	N	17
276	**Waffles**	580	190	260	21	370	1.4	0.11	0.8	N	580	0.30	5	34
277	**Welsh cheesecakes**	400	89	74	11	140	1.2	0.12	0.5	N	470	N	3	17

Buns and pastries *continued*

No. 11-	Food	Retinol equiv µg	Vitamin D µg	Vitamin E mg	Thiamin mg	Ribo-flavin mg	Niacin mg	Trypt 60 mg	Vitamin B6 mg	Vitamin B12 µg	Folate µg	Panto-thenate µg	Biotin µg	Vitamin C mg
264	**Scones,** cheese	170	1.08	1.30	0.14	0.12	0.9	2.2	0.09	Tr	9	0.2	2	Tr
265	fruit	N	N	N	0.24	0.10	1.2	1.5	0.05	Tr	6	N	N	Tr
266	plain	140	1.22	1.44	0.16	0.07	1.0	1.5	0.09	Tr	8	0.2	1	Tr
267	potato	130	1.28	1.45	0.14	0.04	0.8	1.1	0.10	Tr	9	0.2	1	2
268	wholemeal	130	1.15	1.99	0.22	0.10	3.1	1.8	0.23	Tr	18	0.4	5	Tr
269	wholemeal, fruit	120	1.03	1.86	0.20	0.09	2.9	1.6	0.23	Tr	17	0.4	5	Tr
270	**Scotch pancakes**	120	0.92	1.10	0.13	0.09	0.8	1.2	0.08	Tr	6	0.3	1	Tr
271	**Sevyiaan**	290	0.24	N	0.07	0.10	0.6	0.9	0.08	Tr	8	0.2	1	Tr
272	**Strawberry tartlets**	2	0	0.60	0.06	0.02	0.6	0.5	0.05	0	3	0.2	1	26
273	**Teacakes** *fresh*	N	0	N	0.21	0.15	1.8	1.6	0.05	Tr	36	N	N	Tr
274	*toasted*	N	0	N	0.20	0.17	2.0	1.8	0.06	Tr	40	N	N	Tr
275	**Vanilla slices**	16	0.10	0.95	0.08	0.09	0.4	1.0	0.05	Tr	7	0.3	3	1
276	**Waffles**	200	0.34	0.77	0.14	0.18	0.7	2.0	0.10	1	11	0.5	6	1
277	**Welsh cheesecakes**	75	0.66	1.84	0.10	0.06	0.6	1.2	0.05	Tr	7	0.3	3	1

Puddings

Composition of food per 100g

No. 11-	Food	Description and main data sources	Water g	Total nitrogen g	Protein g	Fat g	Carbohydrate g	Energy value kcal	kJ
278	**Apple pie** *one crust*	Recipe	57.8	0.32	1.8	8.0	31.4	197	827
279	*pastry top and bottom*	Recipe	47.2	0.50	2.9	13.3	35.8	266	1115
280	*wholemeal one crust*	Recipe	57.8	0.41	2.4	8.1	29.0	191	805
281	*wholemeal pastry top and bottom*	Recipe. Ref. 7	47.3	0.65	3.8	13.6	31.9	257	1079
282	**Arctic roll**	10 samples, 2 brands	51.3	0.66	4.1	6.6	33.3	200	847
283	**Bakewell tart**	Recipe. Ref. 7	17.9	1.12	6.3	29.7	43.5	456	1903
284	**Blackcurrant pie** *pastry top and bottom*	Recipe. Ref. 7	42.3	0.55	3.1	13.3	34.5	262	1099
285	*wholemeal pastry top and bottom*	Recipe	42.3	0.70	4.1	13.6	30.6	254	1063
286	**Bread and butter pudding**	Recipe	66.7	1.00	6.2	7.8	17.5	160	673
287	**Bread pudding**	Recipe	29.3	0.98	5.9	9.6	49.7	297	1252
288	**Cheesecake**	Recipe	35.3	0.88	5.4	32.4	25.4	411	1706
289	*frozen*	10 samples, assorted flavours, fruit topping	44.0	0.91	5.7	10.6	32.8	241	1013
290	**Christmas pudding**	Recipe	30.4	0.79	4.6	9.7	49.5	291	1227
291	*retail*	10 samples, 4 brands	(23.6)	0.53	3.0	11.8	56.3	329	1388
292	**Crumble** *with pie filling*	Recipe	54.7	0.33	1.9	7.3	36.3	210	883
293	apple	Recipe	53.9	0.31	1.8	6.9	36.6	207	872
294	fruit	Recipe. Apple, gooseberry, plum, rhubarb	54.8	0.35	2.0	6.9	34.0	198	835
295	fruit, wholemeal	Recipe. Apple, gooseberry, plum, rhubarb	54.8	0.44	2.6	7.1	31.7	193	813

No. 11-	Food	Starch	Total sugars	Individual sugars Gluc	Fruct	Sucr	Malt	Lact	Dietary fibre Southgate method	Englyst method	Fibre fractions Cellulose	Non-cellulosic polysaccharide Soluble	Insoluble	Lignin	Resistant starch
278	**Apple pie** *one crust*	13.4	18.0	1.4	3.9	12.7	Tr	0	2.0	1.6	0.4	0.9	0.5	N	N
279	*pastry top and bottom*	22.2	13.7	1.2	3.0	9.5	0.1	0	2.1	1.7	0.3	0.9	0.6	N	N
280	*wholemeal one crust*	10.9	18.1	1.4	3.9	12.8	0	0	2.9	2.6	0.6	1.0	1.2	N	N
281	*wholemeal pastry top and bottom*	18.0	13.8	1.1	3.0	9.7	0	0	3.5	3.4	0.7	1.1	1.8	N	N
282	**Arctic roll**	8.0	25.3	2.6	1.1	16.9	1.7	2.9	0.8	N	N	N	N	N	N
283	**Bakewell tart**	20.7	22.8	4.0	2.4	15.3	1.1	0	2.7	1.9	0.3	0.5	1.0	N	N
284	**Blackcurrant pie** *pastry top and bottom*	22.0	12.5	1.1	1.3	9.4	0.8	0	4.8	2.6	0.4	1.2	1.0	N	N
285	*wholemeal pastry top and bottom*	17.8	12.7	1.1	1.3	9.6	0.8	0	6.2	4.3	0.8	1.3'	2.2	N	N
286	**Bread and butter pudding**	5.2	12.4	1.8	1.7	4.8	Tr	4.1	0.2	0.3	Tr	0.1	0.1	N	N
287	**Bread pudding**	16.6	33.1	8.9	7.8	12.8	1.7	1.9	3.0	1.2	0.4	0.6	0.3	0.6	N
288	**Cheesecake**	11.4	14.0	0.1	0.1	13.4	0	0.4	0.7	0.4	0.1	0.2	0.1	Tr	N
289	*frozen*	10.8	22.0	2.7	1.4	16.0	0.4	1.7	0.9	N	N	N	N	N	N
290	**Christmas pudding**	15.2	34.3	11.4	10.8	11.3	0.9	0	2.7	1.3	0.3	0.6	0.4	N	N
291	*retail*	10.1	46.2	20.3	20.8	3.5	1.5	0.1	3.4	1.7	0.3	N	N	1.0	0.3
292	**Crumble** *with pie filling*	16.8	19.6	3.7	3.9	11.9	Tr	0	1.7	1.2	0.2	0.6	0.4	N	N
293	*apple*	13.0	23.6	1.4	4.1	18.1	Tr	0	2.1	1.6	0.4	0.9	0.5	N	N
294	*fruit*	12.7	21.3	1.5	1.8	18.1	Tr	0	2.2	1.7	0.4	0.8	0.5	N	N
295	*fruit, wholemeal*	10.3	21.4	1.5	1.8	18.2	0	0	3.0	2.7	0.6	0.9	1.2	N	N

Puddings

Inorganic constituents per 100g

No. 11-	Food	Na	K	Ca	Mg	P	Fe	Cu	Zn	S	Cl	Mn	Se	I
							mg						µg	
278	Apple pie one crust	120	100	27	6	30	0.5	0.10	0.2	N	200	0.17	1	5
279	pastry top and bottom	200	100	43	8	40	0.7	0.10	0.3	N	330	0.22	1	7
280	wholemeal one crust	100	140	10	23	66	0.9	0.15	0.6	N	170	0.61	9	N
281	wholemeal pastry top and bottom	170	160	14	37	100	1.3	0.18	0.9	N	270	0.95	15	N
282	Arctic roll	150	140	90	11	120	0.7	0.12	0.4	65	140	0.09	N	23
283	Bakewell tart	290	180	80	42	110	1.5	0.10	0.8	N	460	N	3	17
284	Blackcurrant pie pastry top and bottom	200	220	70	15	53	1.2	0.12	N	N	330	0.41	1	N
285	wholemeal pastry top and bottom	170	280	40	44	110	1.8	0.20	N	N	280	1.15	15	N
286	Bread and butter pudding	150	200	130	16	130	0.7	0.08	0.7	70	240	0.09	6	N
287	Bread pudding	310	310	120	24	110	1.6	0.20	0.6	65	450	0.42	11	N
288	Cheesecake	310	110	64	9	95	0.9	0.11	0.6	N	430	0.08	N	N
289	frozen	160	130	68	10	93	0.5	0.06	0.4	55	220	0.18	N	25
290	Christmas pudding	200	350	79	27	76	1.5	0.22	0.5	48	300	0.43	7	N
291	retail	170	340	35	18	92	1.2	0.14	0.7	490	180	0.51	N	N
292	Crumble with pie filling	100	85	46	7	31	1.4	0.05	0.1	N	150	0.20	1	7
293	apple	68	110	29	5	30	0.6	0.10	0.2	N	120	0.17	1	6
294	fruit	68	190	49	9	33	0.6	0.10	0.2	N	130	0.17	1	6
295	fruit, wholemeal	68	220	32	25	68	0.9	0.15	0.6	6	120	0.59	9	N

No. 11-	Food	Retinol equiv µg	Vitamin D µg	Vitamin E mg	Thiamin mg	Ribo-flavin mg	Niacin mg	Trypt 60 mg	Vitamin B6 mg	Vitamin B12 µg	Folate µg	Panto-thenate µg	Biotin µg	Vitamin C mg
278	**Apple pie** *one crust*	37	0.34	0.52	0.06	0.02	0.3	0.3	0.03	0	4	0.1	Tr	7
279	*pastry top and bottom*	60	0.57	0.76	0.08	0.02	0.5	0.5	0.04	0	4	0.1	Tr	5
280	*wholemeal one crust*	37	0.34	0.71	0.08	0.02	1.0	0.4	0.08	0	7	0.1	1	7
281	*wholemeal pastry top and bottom*	60	0.57	1.08	0.12	0.03	1.6	0.7	0.12	0	9	0.2	2	5
282	**Arctic roll**	N	N	N	0.07	0.11	0.2	0.8	0.06	Tr	12	N	N	0
283	**Bakewell tart**	170	1.64	4.29	0.09	0.15	0.7	1.3	0.05	Tr	12	0.3	3	1
284	**Blackcurrant pie** *pastry top and bottom*	73	0.57	1.14	0.08	0.03	0.6	0.6	0.06	0	N	0.2	1	72
285	*wholemeal pastry top and bottom*	73	0.57	1.46	0.11	0.05	1.7	0.8	0.14	0	N	0.3	3	72
286	**Bread and butter pudding**	110	0.27	0.44	0.07	0.21	0.3	1.6	0.07	1	9	0.5	7	1
287	**Bread pudding**	110	0.16	0.38	0.10	0.12	0.8	1.3	0.09	Tr	8	0.3	4	Tr
288	**Cheesecake**	350	0.94	N	0.04	0.13	0.2	1.3	0.04	Tr	7	N	N	1
289	*frozen*	N	N	N	0.04	0.16	0.3	1.4	0.02	Tr	7	N	N	0
290	**Christmas pudding**	25	0.12	0.64	0.08	0.08	0.7	1.0	0.08	Tr	8	0.2	4	0
291	*retail*	N	N	N	Tr	0.03	0.4	0.6	0.07	Tr	9	N	N	Tr
292	**Crumble** *with pie filling*	71	0.69	0.75	0.05	0.01	0.4	0.4	0.02	0	3	Tr	Tr	2
293	*apple*	69	0.66	0.85	0.06	0.02	0.3	0.3	0.03	0	3	0.1	Tr	5
294	*fruit*	80	0.66	0.98	0.05	0.02	0.5	0.4	0.03	0	3	0.1	Tr	3
295	*fruit, wholemeal*	80	0.66	1.16	0.07	0.03	1.1	0.5	0.08	0	6	0.2	1	3

Puddings *continued*

Composition of food per 100g

No. 11-	Food	Description and main data sources	Water g	Total nitrogen g	Protein g	Fat g	Carbohydrate g	Energy value kcal	kJ
296	**Custard** *made up with* *milk*	Recipe	75.5	0.58	3.7	4.5	16.3	117	492
297	*skimmed milk*	Recipe	79.1	0.60	3.8	0.1	16.6	79	336
298	*canned*	10 samples, 3 brands	77.2	0.42	2.6	3.0	15.4	95	401
299	*confectioners'*	Recipe	63.5	1.03	6.4	5.7	24.2	168	707
300	**Dream Topping**	10 samples of the same brand (Birds)	1.4	0.95	6.0	50.4	39.8	626	2603
301	*made up with milk*	Recipe	70.1	0.59	3.8	13.4	11.8	180	750
302	*made up with* *skimmed milk*	Recipe	72.6	0.61	3.9	10.4	12.0	154	643
303	**Eve's pudding**	Recipe. Ref. 7	52.1	0.58	3.5	13.1	28.9	241	1009
304	**Flan,** *pastry with fruit*	Recipe. Apple, gooseberry, plum, rhubarb	73.5	0.24	1.4	4.4	19.3	118	496
305	*sponge with fruit*	Recipe. Apple, gooseberry, plum, rhubarb	71.3	0.45	2.8	1.5	23.3	112	476
306	**Flan case,** *pastry*	Recipe	3.1	1.22	7.1	33.6	56.7	544	2270
307	*sponge*	Recipe	31.3	1.59	9.8	6.1	53.6	295	1249
308	**Fruit pie** *one crust*	Recipe. Apple, gooseberry, plum, rhubarb	59.0	0.35	2.0	7.9	28.7	186	784
309	*pastry top and bottom*	Recipe. Ref. 7	47.9	0.53	3.0	13.3	34.0	260	1089
310	*individual*	10 pies, as purchased, 3 brands; apple, blackcurrant, blackberry, apricot	22.9	0.75	4.3	15.5	56.7	369	1554
311	*wholemeal one crust*	Recipe. Ref. 7. Apple, gooseberry, plum, rhubarb	58.6	0.45	2.6	8.1	26.6	183	770
312	*wholemeal pastry* *top and bottom*	Recipe. Ref. 7. Apple, gooseberry, plum, rhubarb	47.9	0.68	4.0	13.6	30.0	251	1052

No. 11-	Food	Starch	Total sugars	Individual sugars					Dietary fibre		Fibre fractions				
												Non-cellulosic polysaccharide			
				Gluc	Fruct	Sucr	Malt	Lact	Southgate method	Englyst method	Cellulose	Soluble	Insoluble	Lignin	Resistant starch
296	**Custard** *made up with milk*	5.1	11.2	Tr	0	5.9	0	5.3	N	Tr	Tr	Tr	Tr	0	N
297	*made up with skimmed milk*	5.1	11.4	Tr	0	5.9	0	5.5	N	Tr	Tr	Tr	Tr	0	N
298	*canned*	3.1	12.3	Tr	Tr	7.9	0.3	4.0	0.1	Tr	Tr	Tr	Tr	(0)	N
299	*confectioners'*	5.4	18.7	Tr	Tr	15.0	Tr	3.6	0.3	0.2	Tr	0.1	0.1	0	N
300	**Dream Topping**	9.9	29.9	Tr	Tr	21.5	1.2	7.0	0.1	Tr	Tr	0	0	(0)	Tr
301	*made up with milk*	2.0	9.8	Tr	Tr	4.4	0.2	5.1	Tr	Tr	Tr	0	0	(0)	Tr
302	*made up with skimmed milk*	2.0	9.9	Tr	Tr	4.4	0.2	5.3	Tr	Tr	Tr	0	0	(0)	Tr
303	**Eve's pudding**	11.0	17.9	1.1	1.4	15.4	Tr	0	1.7	1.4	0.3	0.6	0.4	N	N
304	**Flan,** *pastry with fruit*	7.9	11.5	3.0	3.3	5.2	Tr	0	0.7	0.7	0.1	0.4	0.2	N	N
305	*sponge with fruit*	6.4	17.0	2.7	2.8	11.5	Tr	0	0.6	0.6	0.1	0.3	0.2	N	N
306	**Flan case,** *pastry*	44.3	12.4	0.3	0.3	11.6	0.1	0	2.1	1.8	0.1	0.9	0.9	Tr	N
307	*sponge*	18.8	34.8	0.2	0.2	34.5	Tr	0	0.9	0.8	Tr	0.4	0.4	Tr	N
308	**Fruit pie** *one crust*	13.0	15.6	1.4	1.7	12.5	Tr	0	2.1	1.7	0.4	0.8	0.5	N	N
309	*pastry top and bottom*	22.0	12.0	1.1	1.3	9.5	0.1	0	2.2	1.8	0.3	0.8	0.6	N	N
310	*individual*	25.8	30.9	5.7	2.8	21.5	0.9	0	2.3	N	N	N	N	N	N
311	*wholemeal one crust*	10.7	15.9	1.4	1.7	12.8	0	0	3.0	2.7	0.6	0.9	1.2	N	N
312	*wholemeal pastry top and bottom*	17.9	12.2	1.1	1.3	9.7	0	0	3.6	3.5	0.7	1.0	1.8	N	N

Puddings continued

Inorganic constituents per 100g

No. 11-	Food	Na	K	Ca	Mg	P	Fe	Cu	Zn	S	Cl	Mn	Se	I
						mg							µg	
296	**Custard** made up with milk	81	170	130	13	110	0.1	Tr	0.4	35	140	N	N	N
297	made up with skimmed milk	81	170	130	14	110	0.1	0.02	0.4	35	140	N	N	N
298	canned	67	130	100	8	87	0.2	0.02	0.3	32	75	Tr	N	N
299	confectioners'	79	160	110	13	130	0.7	0.03	0.7	70	120	0.05	4	32
300	**Dream Topping**	130	63	23	6	95	0.5	0.13	0.3	78	15	0.04	N	Tr
301	made up with milk	70	130	95	10	93	0.1	0.03	0.3	40	79	0.01	N	17
302	made up with skimmed milk	70	130	95	11	93	0.1	0.03	0.3	40	79	0.01	N	17
303	**Eve's pudding**	170	160	52	9	85	0.8	0.10	0.4	N	220	0.14	3	14
304	**Flan**, pastry with fruit	60	120	22	8	28	0.5	0.03	0.2	N	91	0.08	1	4
305	sponge with fruit	27	120	25	9	45	0.6	0.03	0.3	27	32	0.07	2	9
306	**Flan case**, pastry	400	100	96	14	120	1.9	0.13	0.8	N	660	0.35	4	19
307	sponge	83	110	70	12	140	1.7	0.09	1.0	N	110	0.17	10	34
308	**Fruit pie** one crust	120	180	48	9	33	0.5	0.10	0.2	N	210	0.17	1	5
309	pastry top and bottom	200	160	59	11	43	0.8	0.10	0.2	N	340	0.22	1	7
310	individual	210	120	51	12	64	1.2	0.10	0.5	N	260	2.00	N	N
311	wholemeal one crust	100	210	31	26	69	0.9	0.15	0.6	N	180	0.61	9	12
312	wholemeal pastry top and bottom	170	210	29	39	100	1.3	0.19	0.9	N	280	0.95	15	19

Puddings continued

No. 11-	Food	Retinol equiv µg	Vitamin D µg	Vitamin E mg	Thiamin mg	Ribo-flavin mg	Niacin mg	Trypt 60 mg	Vitamin B_6 mg	Vitamin B_{12} µg	Folate µg	Panto-thenate µg	Biotin µg	Vitamin C mg
296	**Custard** *made up with milk*	64	0.03	0.12	0.05	0.18	0.1	0.9	0.06	1	6	0.4	2	1
297	*made up with skimmed milk*	6	0.01	0	0.05	0.18	0.1	0.9	0.06	1	6	0.4	2	1
298	*canned*	N	N	N	0.04	0.10	Tr	0.6	0.03	Tr	2	N	N	0
299	*confectioners'*	93	0.33	0.51	0.08	0.23	0.2	1.7	0.08	1	15	0.7	8	1
300	**Dream Topping**	N[a]	0	N	0.02	0.25	Tr	1.1	0.02	1	Tr	N	N	0
301	*made up with milk*	N	Tr	N	0.04	0.19	0.1	0.8	0.05	Tr	5	N	N	Tr
302	*made up with skimmed milk*	N	Tr	N	0.04	0.19	0.1	0.8	0.05	Tr	5	N	N	Tr
303	**Eve's pudding**	150	1.30	1.61	0.05	0.07	0.4	0.9	0.04	Tr	6	0.3	4	4
304	**Flan**, *pastry with fruit*	55	0.42	0.51	0.06	0.02	0.4	0.3	0.04	Tr	9	0.1	1	10
305	*sponge with fruit*	36	0.17	0.28	0.06	0.07	0.4	0.7	0.05	Tr	10	0.3	4	9
306	**Flan case**, *pastry*	350	3.26	3.58	0.16	0.06	0.9	1.6	0.09	1	13	0.5	7	0
307	*sponge*	110	0.70	1.01	0.10	0.24	0.4	2.7	0.08	1	18	0.9	15	0
308	**Fruit pie** *one crust*	47	0.34	0.65	0.05	0.02	0.5	0.4	0.03	0	3	0.1	Tr	5
309	*pastry top and bottom*	67	0.57	0.85	0.08	0.02	0.6	0.6	0.04	0	4	0.1	Tr	3
310	*individual*	Tr	0	N	0.05	0.02	0.4	0.9	N	0	N	N	N	Tr
311	*wholemeal one crust*	47	0.34	0.84	0.08	0.03	1.1	0.5	0.08	0	6	0.2	1	5
312	*wholemeal pastry top and bottom*	67	0.57	1.17	0.11	0.03	1.7	0.8	0.12	0	9	0.2	2	4

[a] β-Carotene is added as a colouring agent

Puddings *continued*

No. 11-	Food	Description and main data sources	Water g	Total nitrogen g	Protein g	Fat g	Carbohydrate g	Energy value kcal	kJ
313	**Instant dessert powder**	10 samples, 2 types, different flavours	1.0	0.39	2.4	17.3	60.1	391	1643
314	*made up with milk*	Recipe	72.1	0.48	3.1	6.3	14.6	110	465
315	*made up with skimmed milk*	Recipe	74.8	0.50	3.1	3.2	14.8	83	354
316	**Lemon meringue pie**	Recipe	35.2	0.75	4.5	14.4	45.9	319	1342
317	**Milk puddings**	E.g. rice, sago, semolina, tapioca; recipe	72.4	0.62	3.9	4.3	19.7	128	540
318	*made with skimmed milk*	Recipe	75.9	0.64	4.0	0.2	19.9	92	395
319	**Pancakes,** *sweet*	Recipe	43.4	0.97	5.9	16.2	35.0	301	1260
320	*sweet made with skimmed milk*	Recipe	45.5	0.98	6.0	13.8	35.1	280	1175
321	**Pie** *with pie filling*	Recipe	47.5	0.55	3.2	14.5	34.6	273	1145
322	*wholemeal with pie filling*	Recipe	47.6	0.72	4.2	14.8	30.3	264	1105
323	**Queen of puddings**	Recipe	54.2	0.77	4.8	7.8	33.1	213	898
324	**Rice pudding** *canned*	10 cans, 4 brands	77.6	0.53	3.4	2.5	14.0	89	374
325	**Sponge pudding**	Recipe	32.8	0.97	5.8	16.3	45.3	340	1426
326	*with dried fruit*	Recipe. Ref. 7	30.6	0.90	5.4	14.3	48.1	331	1392
327	*with jam or syrup*	Recipe. Ref. 7	31.8	0.86	5.1	14.4	48.7	333	1398
328	*canned*	10 assorted samples of the same brand (Heinz)	35.3	0.54	3.1	11.4	45.4	285	1201
329	**Spotted dick**	Recipe	34.4	0.72	4.2	16.7	42.7	327	1373

Puddings *continued*

Carbohydrate fractions, g per 100g

No. 11-	Food	Starch	Total sugars	Individual sugars Gluc	Fruct	Sucr	Malt	Lact	Dietary fibre Southgate method	Englyst method	Fibre fractions Cellulose	Non-cellulosic polysaccharide Soluble	Insoluble	Lignin	Resistant starch
313	**Instant dessert powder**	19.4	40.7	Tr	Tr	38.3	0	2.2	1.0	N	0.1	N	N	0.1	N
314	*made up with milk*	3.5	11.1	Tr	Tr	6.9	0	4.2	0.2	N	Tr	N	N	Tr	N
315	*made up with skimmed milk*	3.5	11.3	Tr	Tr	6.9	0	4.3	0.2	N	Tr	N	N	Tr	N
316	**Lemon meringue pie**	21.1	24.8	0.2	0.3	24.3	Tr	0	0.8	0.7	Tr	0.3	0.3	Tr	N
317	**Milk puddings**	9.3	10.5	Tr	Tr	5.5	0	5.0	0.2	0.1	N	N	N	N	N
318	*made with skimmed milk*	9.3	10.7	Tr	Tr	5.5	0	5.2	0.2	0.1	N	N	N	N	N
319	**Pancakes**, sweet	18.8	16.2	0.1	0.1	13.0	Tr	2.9	0.9	0.8	Tr	0.4	0.4	Tr	N
320	*sweet made with skimmed milk*	18.8	16.4	0.2	0.2	13.0	Tr	3.0	0.9	0.8	Tr	0.4	0.4	Tr	N
321	**Pie with pie filling**	26.5	8.1	2.9	3.0	2.1	0.1	0	2.0	1.5	N	0.7	0.6	N	N
322	*wholemeal with pie filling*	22.0	8.3	2.9	3.1	2.3	0	0	3.5	3.4	0.6	0.9	1.9	N	N
323	**Queen of puddings**	4.4	28.7	2.7	1.5	21.3	0.8	2.2	0.4	0.2	Tr	0.2	0.1	0.1	N
324	**Rice pudding** canned	5.8	8.2[a]	Tr	0	5.0	0	3.2	N	0.2	Tr	N	N	N	Tr
325	**Sponge pudding**	26.4	18.9	0.1	0.1	18.0	0.1	0.5	1.2	1.1	Tr	0.5	0.5	Tr	N
326	*with dried fruit*	23.2	24.9	4.3	4.3	15.8	0.1	0.4	1.6	1.2	0.1	0.6	0.5	N	N
327	*with jam or treacle*	23.2	25.5	1.8	1.1	21.6	0.6	0.4	1.1	1.0	Tr	0.5	0.4	Tr	N
328	*canned*	19.6	25.8	4.6	3.8	14.9	2.0	0.6	2.4	0.8	Tr	N	N	0.3	0.3
329	**Spotted dick**	23.9	18.8	N	N	N	N	N	1.5	1.0	0.1	0.5	0.4	N	N

[a] Low calorie varieties contain approximately 3.1g sugar per 100g

Puddings *continued*

No. 11-	Food	Na	K	Ca	Mg	P	Fe	Cu	Zn	S	Cl	Mn	Se	I
						mg							µg	
313	**Instant dessert powder**	1100	64	20	11	650	0.5	0.20	0.4	47	45	0.11	N	Tr
314	*made up with milk*	240	130	97	11	190	0.1	0.04	0.4	33	86	0.02	N	18
315	*made up with skimmed milk*	240	130	97	12	190	0.1	0.04	0.4	33	86	0.02	N	18
316	**Lemon meringue pie**	200	82	45	9	66	0.9	0.08	0.5	N	320	0.14	4	15
317	**Milk puddings**	60	160	120	13	110	0.1	0.01	0.5	37	110	N	N	N
318	*made with skimmed milk*	60	160	120	14	110	0.1	0.02	0.5	37	110	N	N	N
319	**Pancakes,** sweet	53	150	110	14	110	0.8	0.05	0.6	N	100	0.15	3	24
320	*sweet made with skimmed milk*	53	150	110	14	110	0.8	0.06	0.6	N	100	0.15	3	24
321	**Pie** with pie filling	240	91	60	10	44	1.4	0.06	0.2	N	380	0.26	1	8
322	*wholemeal with pie filling*	210	150	28	41	110	2.0	0.16	0.9	N	320	1.06	17	22
323	**Queen of puddings**	140	110	79	11	93	0.7	0.05	0.5	57	220	N	5	24
324	**Rice pudding** canned	50	140	93	11	80	0.2	0.03	0.4	N	95	N	N	N
325	**Sponge pudding**	310	89	84	10	180	1.1	0.10	0.6	N	270	0.21	4	19
326	*with dried fruit*	270	180	82	13	170	1.1	0.15	0.5	N	240	0.23	3	N
327	*with jam or treacle*	290	99	76	10	160	1.1	0.10	0.5	N	240	0.18	3	17
328	*canned*	340	160	50	13	170	1.2	0.31	0.4	34	220	0.21	N	4
329	**Spotted dick**	390	150	96	14	170	0.7	0.15	0.4	N	390	0.24	5	10

No. 11-	Food	Retinol equiv µg	Vitamin D µg	Vitamin E mg	Thiamin mg	Ribo-flavin mg	Niacin mg	Trypt 60 mg	Vitamin B6 mg	Vitamin B12 µg	Folate µg	Panto-thenate µg	Biotin µg	Vitamin C mg
313	**Instant dessert powder**	N	N	N	Tr	0.01	Tr	0.5	Tr	Tr	Tr	N	N	0
314	*made up with milk*	N	N	N	0.04	0.14	0.1	0.7	0.05	Tr	5	N	N	1
315	*made up with skimmed milk*	N	N	N	0.04	0.14	0.1	0.7	0.05	Tr	5	N	N	1
316	**Lemon meringue pie**	100	0.88	1.03	0.07	0.08	0.4	1.1	0.05	Tr	8	0.3	5	3
317	**Milk puddings**	60	0.03	0.11	0.04	0.16	0.1	0.9	0.06	Tr	4	0.3	2	1
318	*made with skimmed milk*	5	0.01	Tr	0.04	0.16	0.1	0.9	0.06	Tr	4	0.3	2	1
319	**Pancakes,** sweet	58	0.17	0.33	0.10	0.17	0.5	1.4	0.09	1	8	0.5	5	1
320	*sweet made with skimmed milk*	27	0.15	0.27	0.10	0.17	0.5	1.4	0.09	1	8	0.5	5	1
321	**Pie** with pie filling	64	0.62	0.72	0.08	0.01	0.6	0.6	0.04	0	4	0.1	Tr	3
322	*wholemeal with pie filling*	64	0.62	1.07	0.11	0.03	1.8	0.8	0.12	0	10	0.2	2	3
323	**Queen of puddings**	100	0.28	0.44	0.05	0.15	0.2	1.2	0.04	1	7	0.4	6	1
324	**Rice pudding** canned	N	N	N	0.03	0.14	0.2	0.7	0.02	Tr	N	N	N	0
325	**Sponge pudding**	170	1.56	1.75	0.09	0.09	0.6	1.4	0.06	Tr	8	0.3	5	Tr
326	*with dried fruit*	150	1.37	1.57	0.09	0.08	0.6	1.2	0.07	Tr	8	0.3	5	Tr
327	*with jam or treacle*	150	1.37	1.54	0.08	0.08	0.5	1.2	0.05	Tr	7	0.3	4	Tr
328	*canned*	N	N	N	0.05	0.16	0.4	0.6	0.09	Tr	3	0.2	1	0
329	**Spotted dick**	27	0.01	0.08	0.09	0.06	0.7	0.9	0.06	Tr	5	0.2	1	Tr

Puddings *continued*

Composition of food per 100g

No. 11-	Food	Description and main data sources	Water g	Total nitrogen g	Protein g	Fat g	Carbohydrate g	Energy value kcal	kJ
330	**Suet pudding**	Recipe	36.0	0.75	4.4	18.3	40.5	335	1401
331	**Treacle tart**	Recipe	21.4	0.64	3.7	14.1	60.4	368	1550
332	**Trifle**	Recipe	67.0	0.59	3.6	6.3	22.4	160	674
333	*frozen*	10 samples, 5 brands, strawberry and raspberry	67.7	0.35	2.2	5.8	20.6	138	581
334	*with Dream Topping*	Recipe	67.9	0.61	3.7	4.7	22.9	148	625
335	*with fresh cream*	10 samples, 7 individual, 2 large	68.1	0.38	2.4	9.2	19.5	166	693

Puddings *continued*

Carbohydrate fractions, g per 100g

| No. 11- | Food | Starch | Total sugars | Individual sugars | | | | | Dietary fibre | | Fibre fractions | | | | |
				Gluc	Fruct	Sucr	Malt	Lact	Southgate method	Englyst method	Cellulose	Non-cellulosic polysaccharide Soluble	Insoluble	Lignin	Resistant starch
330	**Suet pudding**	26.3	14.2	N	N	N	N	N	1.4	0.9	Tr	0.5	0.4	0.1	N
331	**Treacle tart**	26.8	33.6	0.3	0.3	33.0	0.1	0	1.4	1.1	Tr	0.5	0.5	Tr	N
332	**Trifle**	5.6	16.8	2.6	2.0	8.8	0.8	2.6	0.4	0.5	0.1	0.1	0.2	N	N
333	*frozen*	2.7	17.9	4.8	0.6	10.1	1.0	1.4	0.5	(0.5)	(0.1)	(0.1)	(0.2)	N	N
334	*with Dream Topping*	5.7	17.2	2.6	2.0	9.1	0.8	2.7	0.4	0.5	0.1	0.1	0.2	N	N
335	*with fresh cream*	4.5	15.0	1.7	1.4	9.6	0.5	1.8	0.5	(0.5)	(0.1)	(0.1)	(0.2)	N	N

97

No. 11-	Food	Na	K	Ca	Mg	P	mg Fe	Cu	Zn	S	Cl	Mn	μg Se	I
330	**Suet pudding**	420	93	96	13	180	0.7	0.10	0.4	N	430	0.20	6	11
331	**Treacle tart**	360	150	62	13	50	1.4	0.10	0.3	N	430	0.22	4	N
332	**Trifle**	54	150	79	15	85	0.5	0.04	0.4	N	85	0.08	N	87
333	*frozen*	80	29	49	5	52	0.2	0.04	0.2	30	45	0.04	N	(17)
334	*with Dream Topping*	56	150	81	15	87	0.5	0.04	0.4	N	86	0.08	N	88
335	*with fresh cream*	63	84	68	6	63	0.3	0.09	0.3	26	55	0.05	N	17

Puddings continued

No. 11-	Food	Retinol equiv µg	Vitamin D µg	Vitamin E mg	Thiamin mg	Ribo-flavin mg	Niacin mg	Trypt 60 mg	Vitamin B6 mg	Vitamin B12 µg	Folate µg	Panto-thenate mg	Biotin µg	Vitamin C mg
330	**Suet pudding**	30	0.01	0.09	0.09	0.06	0.6	0.9	0.04	Tr	5	0.2	1	Tr
331	**Treacle tart**	60	0.60	0.69	0.08	0.01	0.6	0.7	0.04	0	4	0.1	Tr	0
332	**Trifle**	75	0.12	0.44	0.07	0.13	0.3	0.9	0.06	Tr	9	0.3	3	4
333	*frozen*	(75)	(0.12)	(0.40)	Tr	0.07	0.1	0.5	0.02	Tr	4	(0.3)	(3)	(4)
334	*with Dream Topping*	47	0.11	0.40	0.07	0.13	0.3	0.9	0.06	Tr	9	0.3	3	4
335	*with fresh cream*	(75)	(0.12)	(0.40)	0.06	0.10	0.1	0.5	Tr	Tr	(9)	(0.3)	(3)	(4)

No. 11-	Food	Description and main data sources	Water g	Total nitrogen g	Protein g	Fat g	Carbohydrate g	Energy value kcal	Energy value kJ
336	**Chevda/chevra/chewra**	Recipe	3.6	3.10	17.5	32.3	35.1	494	2054
337	**Chevra and chana chur**	Equal weights of the 2 varieties	3.8	2.53	15.8	39.2	32.8	539	2244
338	**Chinese flaky pastries**	3 assorted samples with bean and vegetable paste filling	21.2	0.95	5.4	16.4	59.3	392	1647
339	**Couscous**	Doughy paste made from millet, ref. 8	40.0	0.98	5.7	1.0	51.3	227	950
340	**Dumplings**	Recipe	60.5	0.48	2.8	11.7	24.5	208	871
341	**Macaroni** canned in cheese sauce	10 samples of the same brand (Heinz)	74.8	0.79	4.5	6.5	16.4	138	579
342	**Macaroni cheese**	Recipe	67.8	1.14	7.1	10.6	13.3	174	725
343	**Masur**	Savoury snack, 5 assorted samples	1.9	0.78	4.9	53.5	38.2	644	2674
344	**Meat buns**, Chinese	2 assorted samples, barbecued pork	40.4	1.54	8.8	9.5	38.4	265	1116
345	**Pakoras/bhajia**	Recipe	56.9	1.25	7.8	8.2	20.4	180	758
346	**Pancakes**, savoury	Recipe	51.9	1.04	6.3	17.5	24.0	273	1138
347	savoury made with skimmed milk	Recipe	54.3	1.06	6.4	14.7	24.1	249	1039
348	**Pilau rice**	Recipe	57.7	0.48	2.7	11.5	25.7	217	904
349	**Pizza**	Cheese and tomato, recipe	51.7	1.46	9.0	11.8	24.8	235	984
350	frozen	10 samples, 2 brands, cheese and tomato	49.3	1.31	7.5	10.7	32.9	250	1050
351	**Ravioli** canned in tomato sauce	10 samples, 4 brands	79.9	0.53	3.0	2.2	10.3	70	297
352	**Risotto** plain	Recipe	55.1	0.51	3.0	9.3	34.4	224	943

| No. 11- | Food | Starch | Total sugars | Individual sugars | | | | | Dietary fibre | | Fibre fractions | | | | Resistant starch |
				Gluc	Fruct	Sucr	Malt	Lact	Southgate method	Englyst method	Cellulose	Non-cellulosic polysaccharide Soluble	Insoluble	Lignin	
336	**Chevda/chevra/chewra**	31.0	2.0	N	N	N	N	N	6.0	3.8	1.1	1.1	1.6	N	N
337	**Chevra and chana chur**	29.4	3.4	N	N	N	N	N	N	N	N	N	N	N	N
338	**Chinese flaky pastries**	29.2	30.1	N	N	N	N	N	N	N	N	N	N	N	N
339	**Couscous**	N	N	N	N	N	N	N	N	N	N	N	N	N	N
340	**Dumplings**	24.0	0.4	0.1	0.1	0.1	0.1	0	1.0	0.9	Tr	0.4	0.4	Tr	N
341	**Macaroni** canned in cheese sauce	14.5	1.9	0.1	0.1	0.9	0.2	0.6	0.8	0.4	0.1	N	N	Tr	0.1
342	**Macaroni cheese**	11.1	2.1	Tr	0	0.1	0.1	1.9	0.8	0.5	0.1	0.3	0.2	Tr	N
343	**Masur**	5.3	32.9	N	N	N	N	N	N	N	N	N	N	N	N
344	**Meat buns**, Chinese	29.4	9.0	N	N	N	N	N	N	N	N	N	N	N	N
345	**Pakoras/bhajia**	17.9	1.8	0.7	0.5	0.6	0	0	5.6	4.5	N	N	N	N	N
346	**Pancakes**, savoury	20.2	3.8	0.1	0.1	0.1	Tr	3.4	1.0	0.8	Tr	0.4	0.4	Tr	N
347	savoury made with skimmed milk	20.2	3.9	0.1	0.1	0.1	Tr	3.5	1.0	0.8	Tr	0.4	0.4	Tr	N
348	**Pilau rice**	20.6	5.1	2.5	2.4	0.1	0	0	1.6	0.6	0.2	0.1	0.2	N	N
349	**Pizza**	22.6	2.2	0.5	0.6	0.9	0.1	0	1.8	1.5	0.2	0.6	0.6	0.1	N
350	frozen	26.0	6.9	0.6	1.0	0.3	4.8	0.2	1.9	(1.5)	N	N	N	N	N
351	**Ravioli** canned in tomato sauce	8.1	2.2	0.5	0.7	0.7	0.3	0	1.0	0.9	0.1	N	N	N	0.1
352	**Risotto** plain	33.2	1.2	0.6	0.3	0.3	0	0	1.3	0.4	0.2	0.2	0.1	0.2	N

in sects.

Savouries

Inorganic constituents per 100g

No. 11-	Food	Na	K	Ca	Mg	P	Fe	Cu	Zn	S	Cl	Mn	Se	I
							mg						µg	
336	Chevda/chevra/chewra	1000	540	53	110	250	5.1	0.29	2.5	210	1510	1.32	N	N
337	Chevra and chana chur	730	530	45	93	250	4.3	0.35	1.9	N	1000	1.00	N	N
338	Chinese flaky pastries	77	110	59	29	94	3.9	0.18	0.7	N	170	0.74	N	N
339	Couscous	N	N	19	N	240	5.0	N	N	N	N	N	N	N
340	Dumplings	400	44	52	8	120	0.6	0.07	0.2	N	460	0.17	1	4
341	Macaroni canned in cheese sauce	560	66	100	11	140	0.3	0.04	0.1	67	850	0.11	N	N
342	Macaroni cheese	320	89	160	17	140	0.4	0.13	0.8	67	490	0.14	4	17
343	Masur	6	170	37	41	93	2.3	0.27	0.8	N	33	0.41	N	N
344	Meat buns, Chinese	350	120	67	22	110	1.8	0.12	1.2	N	630	0.33	N	N
345	Pakoras/bhajia	360	470	93	59	140	3.4	0.24	1.3	67	530	0.87	N	N
346	Pancakes, savoury	150	160	130	16	120	0.8	0.05	0.6	N	250	0.16	3	26
347	savoury made with skimmed milk	150	160	130	16	120	0.8	0.06	0.6	N	250	0.16	3	26
348	Pilau rice	110	150	19	20	41	0.6	0.13	0.6	34	180	0.40	N	N
349	Pizza	570	150	190	18	160	1.0	0.12	0.9	(92)	940	0.21	4	14
350	frozen	540	170	180	16	130	1.0	0.11	1.0	92	910	0.26	(4)	(14)
351	Ravioli canned in tomato sauce	490	150	16	12	43	0.8	0.08	0.5	44	760	0.16	N	N
352	Risotto plain	410	82	24	16	64	0.3	0.16	0.8	43	640	0.49	4	N

No. 11-	Food	Retinol equiv µg	Vitamin D µg	Vitamin E mg	Thiamin mg	Ribo-flavin mg	Niacin mg	Trypt 60 mg	Vitamin B6 mg	Vitamin B12 µg	Folate µg	Panto-thenate µg	Biotin µg	Vitamin C mg
336	Chevda/chevra/chewra	89	0	6.31	0.39	0.09	6.3	3.5	0.26	0	30	1.1	10	2
337	Chevra and chana chur	Tr	N	N	0.16	0.12	8.2	2.9	N	Tr	8	N	N	N
338	Chinese flaky pastries	N	N	N	0.09	0.02	0.6	1.1	N	Tr	N	N	N	N
339	Couscous	Tr	0	N	0.20	0.06	0.8	1.1	N	0	N	N	N	0
340	Dumplings	9	Tr	0.09	0.05	0.01	0.3	0.6	0.03	Tr	3	0.1	Tr	0
341	Macaroni canned in cheese sauce	N	N	N	0.06	0.06	0.4	0.9	0.08	1	6	N	N	0
342	Macaroni cheese	110	0.37	0.50	0.04	0.12	0.3	1.6	0.04	Tr	5	0.2	1	Tr
343	Masur	140	N	N	0.03	0.17	0.3	0.9	N	Tr	16	N	N	N
344	Meat buns, Chinese	N	N	N	0.21	0.12	1.5	1.8	N	Tr	N	N	N	5
345	Pakoras/bhajia	130	0	N	0.15	0.05	0.6	1.1	0.16	0	50	0.4	N	5
346	Pancakes, savoury	63	0.16	0.34	0.11	0.19	0.5	1.5	0.10	1	8	0.5	5	1
347	savoury made with skimmed milk	26	0.15	0.27	0.11	0.19	0.5	1.5	0.10	1	8	0.5	5	1
348	Pilau rice	0	0	3.03	0.11	0.03	1.1	0.5	0.09	0	6	0.2	1	Tr
349	Pizza	76	0.06	1.27	0.10	0.13	0.9	2.0	0.09	Tr	23	0.3	3	3
350	frozen	N	N	N	0.16	0.14	0.9	2.0	0.13	Tr	20	N	N	N
351	Ravioli canned in tomato sauce	N	0	N	0.05	0.04	0.9	0.6	0.10	Tr	3	N	N	Tr
352	Risotto plain	77	0.77	0.81	0.15	0.01	1.5	0.7	0.12	0	5	0.2	1	Tr

Savouries *continued*

11-353 *to* 11-360

Composition of food per 100g

No. 11-	Food	Description and main data sources	Water g	Total nitrogen g	Protein g	Fat g	Carbohydrate g	Energy value kcal	kJ
353	**Samosa**, meat	Recipe	20.4	0.85	5.1	56.1	17.9	593	2451
354	vegetable	Recipe	31.5	0.52	3.1	41.8	22.3	472	1954
355	**Sev/ganthia**	Savoury Indian snack, recipe	3.8	2.88	18.0	25.3	41.2	453	1899
356	**Spaghetti** *canned in bolognese sauce*	10 samples of the same brand (Heinz)	79.1	0.58	3.3	3.0	12.2	86	362
357	*canned in tomato sauce*	10 samples, 3 brands	81.9	0.33	1.9	0.4	14.1	64	273
358	**Stuffing**, sage and onion	Recipe	56.5	0.87	5.2	14.8	20.4	231	962
359	**Yorkshire pudding**	Recipe	57.4	1.09	6.6	9.9	24.7	208	874
360	*made with skimmed milk*	Recipe	59.6	1.10	6.7	7.3	24.8	185	779

Savouries *continued*

Carbohydrate fractions, g per 100g

No. 11-	Food	Starch	Total sugars	Individual sugars					Dietary fibre		Fibre fractions			Lignin	Resistant starch
									Southgate method	Englyst method		Non-cellulosic polysaccharide			
				Gluc	Fruct	Sucr	Malt	Lact			Cellulose	Soluble	Insoluble		
353	**Samosa**, meat	16.8	1.0	0.3	0.2	0.5	Tr	0	1.9	1.2	N	N	N	N	N
354	vegetable	20.1	1.9	0.6	0.4	0.9	Tr	0	2.4	1.8	0.6	0.5	0.3	N	N
355	**Sev/ganthia**	40.3	0.8	0	0	0.8	0	0	11.1	8.8	2.1	2.7	4.0	N	N
356	**Spaghetti** *canned in bolognese sauce*	9.7	2.5	0.5	0.7	1.0	0.2	Tr	1.4	0.9	N	N	N	0.1	N
357	*canned in tomato sauce*	8.6	5.5	1.0	1.1	2.9	0.4	0	2.8[a]	0.7[a]	0.2	N	N	0.6	0.2
358	**Stuffing**, sage and onion	14.6	5.8	N	N	N	N	N	2.4	1.7	0.5	1.0	0.3	0.1	N
359	**Yorkshire pudding**	21.0	3.7	0.1	0.1	0.1	Tr	3.3	1.0	0.9	Tr	0.4	0.4	Tr	N
360	*made with skimmed milk*	21.0	3.8	0.1	0.1	0.1	Tr	3.4	1.0	0.9	Tr	0.4	0.4	Tr	N

[a]Wholemeal types contain approximately 6.2g Southgate fibre and 2.0g Englyst fibre per 100g

Savouries continued

Inorganic constituents per 100g

No. 11-	Food	Na	K	Ca	Mg	P	Fe	Cu	Zn	S	Cl	Mn	Se	I
						mg							µg	
353	**Samosa**, meat	33	120	34	11	62	0.8	0.07	0.7	N	67	0.22	1	4
354	vegetable	200	200	32	15	47	0.8	0.08	0.3	N	340	0.79	N	4
355	**Sev/ganthia**	610	660	160	110	310	7.6	0.57	2.9	150	950	1.89	N	N
356	**Spaghetti** canned in													
	bolognese sauce	410	140	18	14	45	0.7	0.03	0.4	51	670	0.16	N	N
357	canned in tomato sauce	420	110	12	10	29	0.3	0.06	0.3	29	500	0.13	N	N
358	**Stuffing**, sage and onion	420	150	58	13	77	1.0	0.11	0.6	84	650	0.27	11	15
359	**Yorkshire pudding**	590	160	130	19	120	0.9	0.05	0.6	N	940	0.17	3	27
360	made with skimmed milk	590	160	130	20	120	0.9	0.06	0.6	N	940	0.17	3	27

Savouries *continued*

No. 11-	Food	Retinol equiv µg	Vitamin D µg	Vitamin E mg	Thiamin mg	Riboflavin mg	Niacin mg	Trypt/60 mg	Vitamin B_6 mg	Vitamin B_{12} µg	Folate µg	Pantothenate µg	Biotin µg	Vitamin C mg
353	**Samosa**, meat	28	0.02	11.85	0.09	0.05	1.0	1.1	0.07	Tr	6	0.1	Tr	1
354	vegetable	31	0.01	9.75	0.12	0.02	0.6	0.6	0.09	0	11	0.1	Tr	4
355	**Sev/ganthia**	51	0	N	0.29	0.09	0.9	2.2	0.25	0	74	0.9	N	1
356	**Spaghetti** *canned in bolognese sauce*	N	0	N	0.07	0.03	0.8	0.7	0.10	Tr	8	N	N	0
357	*canned in tomato sauce*	N	0	N	0.07	0.01	0.6	0.4	0.07	Tr	5	Tr	Tr	Tr
358	**Stuffing**, sage and onion	159	1.41	1.48	0.12	0.07	0.8	1.2	0.10	Tr	13	0.3	4	2
359	**Yorkshire pudding**	65	0.19	0.37	0.10	0.16	0.5	1.6	0.07	1	9	0.4	5	1
360	*made with skimmed milk*	30	0.17	0.30	0.10	0.16	0.5	1.6	0.07	1	9	0.4	5	1

REFERENCES TO TABLES

1   Chughtai, M. I. D. and Waheed Khan, A. (1960) *Nutritive value of food-stuffs and planning of satisfactory diets in Pakistan, Part 1. Composition of raw food-stuffs*, Punjab University Press, Lahore

2   Gopalan, C., Rama Sastri, B. V. and Balasubramanian, S. C. (1980) *Nutritive value of Indian foods*, National Institute of Nutrition, Indian Council of Medical Research, Hyderabad

3   Polacchi, W., McHargue, J. S. and Perloff, B. P. (1982) *Food composition tables for the near east.*, Food and Agriculture Organization of the United Nations, Rome

4   Swaminathan, M. (1974) *Essentials of food and nutrition, Volume 1. Fundamental aspects*, Garesh & Co., Madras

5   Watt, B. K. and Merrill, A. L. (1963) *Composition of foods, raw, processed and prepared*, Agriculture Handbook No. 8, US Department of Agriculture, Washington DC

6   Wharton, P. A., Eaton, P. M. and Day, K. C. (1983) Sorrento Asian food tables: food tables, recipes and customs of mothers attending Sorrento Maternity Hospital, Birmingham, England. *Hum. Nutr.: Appl. Nutr.*, **37A**, 378-402

7   Wiles, S. J., Nettleton, P. A., Black. A. E. and Paul, A. A. (1980) The nutrient composition of some cooked dishes eaten in Britain: A supplementary food composition table. *J. Hum. Nutr.*, **34**, 189-223

8   Wu Leung, W. T., Busson, F. and Jardin, C. (1968) *Food composition table for use in Africa*, Food and Agriculture Organization and US Department of Health, Education and Welfare, Bethesda

9   Wu Leung, W. T., Butrum, R. R., Chang, F. H., Narayana Rao, M. and Polacchi, W. (1972) *Food composition table for use in East Asia*, Food and Agriculture Organization and US Department of Health, Education and Welfare, Bethesda

PHYTIC ACID

In contrast to previous editions of *The Composition of Foods* where results were expressed as phytic acid phosphorus, the results below are as grams of phytic acid per 100g food. Figures are from analyses and calculations from recipes, together with calculations from earlier phytic acid phosphorus values on the basis that 1g phytic acid phosphorus is equivalent to 3.55g phytic acid.

		g per 100g			g per 100g
Flours, grains and starches			94	**Vitbe** *average*	0.44
			98	**Wheatgerm bread**	
2	**Barley**, pearl *raw*	0.48		*average*	0.35
3	pearl *boiled*	0.20	99	**White bread** *average*	0.05
4	whole grain *raw*	1.07	105	*fried*	0.05
5	**Bran**, wheat	3.41	106	*toasted*	0.06
8	**Chapati flour**, brown	0.70	113	**Wholemeal bread**	
9	white	0.58		*average*	0.60
18	**Oatmeal**, quick		117	*toasted*	0.70
	cook *raw*	0.96			
19	**Popcorn**, candied	0.21	**Rolls**		
20	plain	0.63			
22	**Rye flour** *whole*	0.92	118	**Brown rolls** *crusty*	0.35
25	**Soya flour** *full fat*	1.50	119	*soft*	0.35
28	**Wheat flour**, brown	0.50	120	**Croissants**	0.09
30	white *breadmaking*	0.18	121	**Hamburger buns**	0.05
31	white *plain*	0.13	122	**Morning rolls**	0.05
32	white *self-raising*	0.18	123	**White rolls** *crusty*	0.05
34	**Wheatgerm**	4.00	124	*soft*	0.05
			125	**Wholemeal rolls**	0.60
Rice					
			Breakfast cereals		
42	**White rice**, easy				
	cook *raw*	0.42	126	**All-Bran**	3.50
45	fried	0.01	130	**Corn Flakes**	0.04
49	polished *raw*	0.42	137	**Muesli** *Swiss style*	0.68
			145	**Ready Brek**	0.84
Breads			146	**Rice Krispies**	0.18
			148	**Shredded Wheat**	0.94
66	**Bannocks** *made*		154	**Weetabix**	0.45
	with beremeal	0.03			
67	*made with wheat flour*	0.10	**Biscuits**		
70	**Brown bread** *average*	0.35			
73	*toasted*	0.45	165	**Brandy snaps**	0.03
75	**Chapatis** *made*		168	**Crispbread**, rye	0.40
	without fat	0.37	169	**Digestive biscuits**,	
79	**Hovis** *average*	0.25		chocolate	0.11
83	*toasted*	0.32	170	plain	0.14
85	**Milk bread**	0.11	171	**Flapjacks**	0.37
86	**Naan bread**	0.11	172	**Gingernut biscuits**	0.09
89	**Paratha**	0.32	173	homemade	0.08
91	**Rye bread**	0.90	179	**Melting moments**	0.04
92	**Soda bread**	0.09	182	**Sandwich biscuits**	0.05
93	**Tortillas** *made*		183	**Semi-sweet biscuits**	0.09
	with wheat flour	0.10	185	**Shortbread**	0.08

Cakes

189	**All-Bran loaf**	0.52
190	**Battenburg cake**	0.03
193	**Cherry cake**	0.03
195	**Chocolate cake**	0.06
196	*with butter icing*	0.05
201	**Fruit cake**, rich	0.02
203	rich, iced	0.02
205	**Gateau**	0.02
206	**Gingerbread**	0.05
208	**Lardy cake**	0.07
210	**Rock cakes**	0.05
211	**Sponge cake**	0.04
212	*fatless*	0.04
214	*with butter icing*	0.03
218	**Welsh cakes**	0.05

Pastry

219	**Cheese pastry** *cooked*	0.06
220	**Choux pastry** *raw*	0.03
221	*cooked*	0.05
222	**Flaky pastry** *raw*	0.06
223	*cooked*	0.08
225	**Shortcrust pastry** *raw*	0.08
226	*cooked*	0.09

Buns and pastries

232	**Chelsea buns/Bath buns**	0.08
233	**Choux buns**	0.03
234	**Cream horns**	0.05
238	**Custard tart** *large*	0.04
245	**Eccles cake**	0.04
246	**Eclairs** *fresh*	0.02
253	**Hot cross buns**	0.08
257	**Jellabi**	0.08
258	**Mince pies,** individual	0.05
259	**Mincemeat tart** *one crust*	0.04
260	**Muffins**	0.11
261	bran	0.68
262	**Pinni**	0.14

263	**Rum baba**	0.02
264	**Scones**, cheese	0.07
266	plain	0.08
270	**Scotch pancakes**	0.06
272	**Strawberry tartlets**	0.03
276	**Waffles**	0.09

Puddings

278	**Apple pie** *one crust*	0.02
279	*pastry top and bottom*	0.04
286	**Bread and butter pudding**	0.01
287	**Bread pudding**	0.02
288	**Cheesecake**	0.02
290	**Christmas pudding**	0.02
299	**Custard,** confectioners'	0.01
306	**Flan case**, pastry	0.08
307	sponge	0.03
316	**Lemon meringue pie**	0.03
319	**Pancakes**, sweet	0.03
320	sweet *made with skimmed milk*	0.03
321	**Pie** *with pie filling*	0.04
325	**Sponge pudding**	0.04
326	*with dried fruit*	0.04
327	*with jam or treacle*	0.04
329	**Spotted dick**	0.03
330	**Suet pudding**	0.03
331	**Treacle tart**	0.04

Savouries

336	**Chevda/chevra/chewra**	0.12
340	**Dumplings**	0.04
346	**Pancakes**, savoury	0.03
347	savoury *made with skimmed milk*	0.03
348	**Pilau rice**	0.11
349	**Pizza**	0.04
352	**Risotto** *plain*	0.17
355	**Sev/ganthia**	0.65
358	**Stuffing**, sage and onion	0.03
359	**Yorkshire pudding**	0.04
360	*made with skimmed milk*	0.04

CHOLESTEROL

A number of sterols are found in cereals either in the free form or esterified with fatty acids. The main sterol in most cereal products, however, is cholesterol derived from the other ingredients in the food.

The table below shows cholesterol alone. The values are derived from two sources: direct analyses of samples and calculation from recipes. In the recipes the cholesterol content will be dependent on the type of fat used. Units are mg/100g food and can be converted into mmol cholesterol by dividing by 386.6.

mg per 100g

Flours, grains and starches

| 19 | **Popcorn**, candied | 18 |

Rice

| 39 | **Savoury rice** raw | 1 |
| 45 | **White rice**, fried | 2 |

Pasta

51	**Lasagna** raw	8
52	boiled	2
55	**Noodles**, egg raw	30
56	egg boiled	6
57	fried	5

Breads

66	**Bannocks** made with beremeal	6
67	made with wheat flour	6
76	**Currant bread**	12
77	toasted	13
84	**Malt bread**	1
85	**Milk bread**	6
86	**Naan bread**	9
89	**Paratha**	38
92	**Soda bread**	6

Rolls

| 120 | **Croissants** | 52 |

Breakfast cereals

| 141 | **Porridge** made with milk | 14 |
| 142 | made with milk and water | 7 |

Biscuits

| 165 | **Brandy snaps** | 58 |
| 170 | **Digestive biscuits**, plain | 22 |

mg per 100g

174	**Homemade biscuits**	
	creaming method	56
175	rubbing-in method	53
176	wholemeal	9
177	**Jaffa cakes**	21
180	**Oatcakes** homemade	8
185	**Shortbread**	74

Cakes

189	**All-Bran loaf**	4
190	**Battenburg cake**	88
191	**Cake mix**	5
192	made up	78
195	**Chocolate cake**	121
196	with butter icing	93
197	**Coconut cake**	78
201	**Fruit cake**, rich	53
202	rich retail	80
203	rich, iced	41
204	wholemeal	57
205	**Gateau**	166
206	**Gingerbread**	59
208	**Lardy cake**	15
210	**Rock cakes**	42
211	**Sponge cake**	128
212	fatless	261
214	with butter icing	90
215	frozen	63
216	**Swiss roll**	188
217	**Swiss rolls**,	
	chocolate, individual	86
218	**Welsh cakes**	71

Pastry

219	**Cheese pastry** cooked	32
220	**Choux pastry** raw	112
221	cooked	173

222	**Flaky pastry** *raw*	12
223	*cooked*	15
224	**Puff pastry** *frozen, raw*	57
225	**Shortcrust pastry** *raw*	11
226	*cooked*	12
227	*frozen, raw*	41
228	**Wholemeal pastry** *raw*	11
229	*cooked*	12

Buns and pastries

232	**Chelsea buns/Bath buns**	37
233	**Choux buns**	158
234	**Cream horns**	38
237	**Currant buns**	17
238	**Custard tart** *large*	57
239	**Custard tarts**, individual	95
240	**Danish pastries**	41
241	**Doughnuts**, custard-filled	41
242	jam	15
243	ring	24
244	ring, iced	18
245	**Eccles cake**	18
246	**Eclairs** *fresh*	105
247	*frozen*	150
249	**Gulab jamen/gulab jambu** *homemade*	19
250	*retail*	37
253	**Hot cross buns**	26
254	**Jam tarts**	6
255	*retail*	42
256	wholemeal	6
260	**Muffins**	41
261	bran	59
263	**Rum baba**	33
264	**Scones**, cheese	16
265	fruit	27
266	plain	5
267	potato	3
268	wholemeal	5
269	wholemeal, fruit	5
270	**Scotch pancakes**	7
271	**Sevyiaan**	77
272	**Strawberry tartlets**	4
273	**Teacakes** *fresh*	18
274	*toasted*	20
275	**Vanilla slices**	42
276	**Waffles**	121
277	**Welsh cheesecakes**	61

Puddings

278	**Apple pie** *one crust*	3
279	*pastry top and bottom*	5
280	wholemeal *one crust*	3
281	wholemeal *pastry top and bottom*	5
282	**Arctic roll**	30
283	**Bakewell tart**	61
286	**Bread and butter pudding**	102
287	**Bread pudding**	59
288	**Cheesecake**	109
289	*frozen*	61
290	**Christmas pudding**	49
291	*retail*	36
296	**Custard** *made up with milk*	16
297	*made up with skimmed milk*	2
298	canned	9
299	confectioners'	126
300	**Dream Topping**	4
301	*made up with milk*	12
302	*made up with skimmed milk*	2
303	**Eve's pudding**	64
304	**Flan**, pastry *with fruit*	16
305	sponge *with fruit*	63
306	**Flan case**, pastry	124
307	sponge	264
308	**Fruit pie** *one crust*	3
309	*pastry top and bottom*	5
311	wholemeal *one crust*	3
312	wholemeal *pastry top and bottom*	5
313	**Instant dessert powder**	1
314	*made up with milk*	12
315	*made up with skimmed milk*	2
316	**Lemon meringue pie**	87
317	**Milk puddings**	15
318	*made with skimmed milk*	2
319	**Pancakes**, sweet	73
320	sweet *made with skimmed milk*	65
323	**Queen of puddings**	102
325	**Sponge pudding**	78
326	*with dried fruit*	69
327	*with jam or treacle*	69
328	*canned*	32
329	**Spotted dick**	14
330	**Suet pudding**	16
331	**Treacle tart**	5

Cholesterol *continued* mg per 100g

332 **Trifle**	50
333 *frozen*	20
334 *with Dream Topping*	44
335 *with fresh cream*	33

Savouries

340 **Dumplings**	8
341 **Macaroni** *canned in cheese sauce*	8
346 **Pancakes**, savoury	73
347 savoury *made with skimmed milk*	64

mg per 100g

349 **Pizza**	16
350 *frozen*	26
351 **Ravioli** *canned in tomato sauce*	6
353 **Samosa**, meat	20
354 vegetable	4
356 **Spaghetti** *canned in bolognese sauce*	4
358 **Stuffing**, sage and onion	63
359 **Yorkshire pudding**	76
360 *made with skimmed milk*	67

RECIPES

- In the recipe calculations, an egg has been assumed to weigh 50g. A level teaspoon refers to a standard 5ml spoon and has been taken to hold 5g salt and 3.5g baking powder, bicarbonate of soda and cream of tartar.
- The fat used in domestic recipes was butter or margarine as indicated. In recipes for commercial products, however, it was taken as unfortified margarine and these products thus contain less vitamin A and vitamin D.
- Where the quantity of vegetable oil is in brackets, this represents a measured amount absorbed during deep fat frying.
- The majority of weight losses were determined experimentally, although some represent estimations from similar foods.

19 Candied popcorn

45ml oil
75g popping corn

Glaze:
45ml water
200g caster sugar
25g butter

Prepare corn as for plain popcorn (No. 20). Heat glaze ingredients until sugar has dissolved, boil to soft ball stage. Add the popped corn and stir until coated.

20 Plain popcorn

45ml oil   75g popping corn

Heat oil gently in saucepan until test corn pops. Remove from heat, add corn, cover and return to heat until all corn has popped.

45 Fried rice

550g boiled rice
168g chopped onion
21g dripping or lard
21g clove garlic

2g salt
pepper
1g spices

Fry onion and garlic until soft. Add boiled rice and seasoning. Fry until fat has been absorbed and rice is fully coated.

Weight loss: 5.6%

57 Fried noodles

450g cooked egg noodles
60ml oil

30g spring onions

Chop onions and stir fry for 30 secs. Add noodles and stir fry for 2 mins. Add salt to taste.

Weight loss: 2%

66 Beremeal bannocks

250g barleymeal
125g plain flour
1 tsp baking powder

1/2 tsp cream of tartar
pinch of salt
188ml buttermilk

Mix dry ingredients and add milk to form a stiff dough. Roll out dough thinly. Cook on a griddle until brown.

Weight loss: 15.8%

67 Flour bannocks

250g plain flour
125ml buttermilk

1 tsp baking powder
1/2 tsp cream of tartar

As for beremeal bannocks.

Weight loss: 15.8%

85 Milk bread

450g strong flour
56g margarine
7g salt

28g fresh yeast
280ml tepid milk

Rub fat into flour. Cream yeast and dissolve salt in the milk, add flour mixture and knead. Leave for 30 mins, knead again and leave for a further 20 mins. Shape and put in tin. After 15 mins glaze and cook at 250°C/mark 8 for 30-40 mins.

Weight loss: 10.5%

86 Naan bread

336g flour
1 tsp salt
112g natural yoghurt
150ml milk

2 tsp sugar
0.5g bakers yeast
56g ghee

Sift flour with salt and mix in yoghurt. Add warmed milk, sugar and yeast. Knead well for about 15 mins then leave for 4-5 hrs. Shape and bake 230°C/mark 8 for 10 mins.

Weight loss: 16.8%

89 Paratha

363g brown flour
245g water
92g butter

Make a dough with the flour and water, chill for 2-3 hrs. Incorporate fat by spreading, folding and rolling as in puff pastry. Fry for about 2-3 mins on each side until crisp.

Weight loss: 17.8%

92 **Soda bread**

500g flour
1 tsp salt
1 tsp bicarbonate of soda

1 tsp cream of tartar
290ml milk

Sift the dry ingredients and quickly knead to a soft dough with the milk. Bake for 35 mins at 220°C/mark 7.

Weight loss: 8.4%

93 **Tortillas**

266g flour
150ml water
1/2 tsp salt

Gradually stir in water to form a firm dough. Roll out thinly. Cook on a griddle for 1 min each side.

Weight loss: 17.3%

120 **Croissants**

450g strong flour
170g unfortified margarine
28g dry yeast
28g lard
1 egg
1 tsp salt
240ml water

Glaze:
30ml water
2.5g caster sugar
1 egg

Blend yeast and water, sift flour, salt and rub in lard. Mix together and knead until smooth. Roll out, dot with margarine and fold into three. Repeat twice, cover and rest for 30 mins. Repeat process a further 3 times, then place in fridge for 1 hr. Roll out, trim, cut and shape into crescents. After 30 mins brush with glaze and bake for 20 mins at 220°C/mark 7.

Weight loss: 15%

141 **Porridge, made with milk**

60g oatmeal
7g salt
Weight loss: 14%

500ml milk

142 **Porridge, made with milk and water**

60g oatmeal
7g salt
Weight loss: 14%

250ml milk
250ml water

143 **Porridge, made with water**

60g oatmeal
7g salt
Weight loss: 14%

500ml water

165 Brandy snaps

56g butter
56g caster sugar
56g golden syrup

56g plain flour
2g ginger
5ml brandy

Melt butter, add sugar with syrup and heat. Stir in sifted flour and ginger. Place teaspoonsful on baking sheets well apart. Bake 180°C/mark 4 for 5-10 mins. Cool slightly and roll each round the greased handle of a wooden spoon.

Weight loss: 10%

171 Flapjacks

120g rolled oats
90g margarine
60g golden syrup

60g brown sugar
2g ginger

Melt fat, add sugar and syrup. Work in the oats. Press into a greased sandwich tin and bake at 170°C/mark 3 for 30 mins.

Weight loss: 5%

173 Gingernut biscuits

112g self-raising flour
1g salt
1/2 tsp bicarbonate of soda
5g ginger

84g golden syrup
56g margarine
10g caster sugar

Sift the flour, baking soda, ginger and salt. Melt the fat and stir in syrup. Stir into the dry ingredients and mix well. Roll into small balls and flatten slightly. Bake for 15 mins at 190°C/mark 5.

Weight loss: 10%

174 Homemade biscuits - creaming method

200g plain flour
1 egg

100g margarine
100g caster sugar

Cream fat and sugar. Mix in egg, then flour and knead the dough lightly until smooth. Roll out thinly, prick and shape. Bake 10-15 mins at 180°C/mark 4.

Weight loss: 10%

175 Homemade biscuits - rubbing in method

224g plain flour
1 tsp baking powder
1/2 tsp salt

112g margarine
1 egg
84g sugar

Sift flour, baking powder and salt. Rub in fat, add sugar and egg, mixing to a stiff dough. Roll out and cut into shapes. Bake 10 mins at 180°C/mark 4.

Weight loss: 10%

176 Homemade wholemeal biscuits

224g wholemeal flour　　　　14g margarine
1/2 tsp salt　　　　　　　　　150ml milk

Mix flour and salt, rub in fat. Mix to a stiff dough with milk. Roll thinly and cut into rounds. Bake for 20 mins at 180°C/mark 4.

Weight loss: 36.7%

177 Jaffa cakes

33.1% baked sponge base
39.6% orange jelly
27.3% plain chocolate

Recipe from FMBRA (Flour Milling and Baking Research Association).

179 Melting moments

112g margarine　　　　　　28g cornflour
28g icing sugar　　　　　　vanilla essence
84g flour　　　　　　　　　28g glacé cherries

Cream margarine, icing sugar and vanilla essence. Sift flour and cornflour, work into fat and sugar. Shape and decorate with cherry pieces. Bake at 190°C/mark 5 for 10 mins.

Weight loss: 10%

180 Oatcakes

28g lard　　　　　　　　　1/2 tsp salt
100ml boiling water　　　　1/4 tsp bicarbonate of soda
224g oatmeal

Mix dry ingredients. Add melted fat with sufficient water to make a stiff paste. Knead, roll out thinly and cut. Cook on a hot griddle or bake at 150°C/mark 2.

Weight loss: 26.8%

185 Shortbread

200g flour　　　　　　　　100g butter
50g caster sugar

Beat the butter and sugar to a cream. Mix in the flour and knead until smooth. Press into a flat tin to about 2cms thick. Bake for about 45 mins at 170°C/mark 3.

Weight loss: 10%

188 Wholemeal crackers/Farmhouse crackers

105g fat　　　　　　　　　630g plain flour
11.9g salt　　　　　　　　210g wholemeal flour
15.4g bakers yeast　　　　2.1g bicarbonate of soda

Recipe from FMBRA.

Weight loss: 11%

189 All-Bran loaf

125g All-Bran
150g sugar
250g milk
125g flour

92g raisins
92g sultanas
91g currants

Mix All-Bran, sugar and dried fruit. Stir in milk and leave to stand for 30 mins. Sieve in the flour, mixing well and pour mixture into a well greased loaf tin. Bake at 180°C/mark 4 for 1 1/4 hrs.

Weight loss: 5%

190 Battenburg cake

100g flour
75g margarine
120g sugar
90g eggs

67.5g water
8g skimmed milk powder
5g baking powder
2g salt

58g marzipan

8g jam

Recipe from FMBRA.

Weight loss: 10% cake

192 Cake mix - made up

205g powder mix
1 egg
85ml water

Add egg and some of the water to the powder, whisk vigorously. Add remaining water, repeat whisking until smooth and creamy. Bake at 190°C/mark 5 for 20-25 mins.

Weight loss: 11.2%

193 Cherry cake

84g margarine
84g caster sugar
2 eggs
126g flour

1g pinch of salt
3/4 tsp baking powder
14g cornflour
168g glacé cherries

Sift flour, salt, cornflour and baking powder. Add cherries. Cream fat and sugar, gradually beat in eggs. Fold in flour until the mixture is even. Bake for 45 mins at 180°C/mark 4.

Weight loss: 12.9%

195/196 Chocolate cake (with butter icing)

150g flour
3 eggs
150g sugar

40g cocoa
150g margarine
5g baking powder

Icing: 85g icing sugar
                85g margarine

Cream fat and sugar. Gradually add beaten egg and beat well. Fold in flour, cocoa and baking powder. Bake for about 20 mins at 190°C/mark 5. Cool and, for No. 196, sandwich with butter icing.

Weight loss: 13.7% for cake

197 Coconut cake

224g plain flour   112g sugar
3 tsp baking powder   56g desiccated coconut
1/8 tsp salt   2 eggs
112g margarine   60ml milk

Sift flour salt and baking powder. Rub in fat, add sugar and coconut. Mix with the egg gradually adding milk. Bake for 1-11/4 hrs at 180°C/mark 4.

Weight loss: 12.9%

198 Crispie cakes

112g plain chocolate
33g crisp rice cereal
33g corn flake type cereal

Melt chocolate in a bowl over hot water. Stir in cereals. Place in cases and allow to cool and set.

Weight loss: 0%

201 Rich fruit cake

200g margarine   250g flour
200g brown sugar   1/4 tsp salt
4 eggs   750g mixed fruit
20g black treacle   150g mixed glacé fruit, chopped
20ml brandy   1 tsp mixed spice

Cream the fat and sugar. Beat the eggs, treacle and brandy. Fold in the sifted flour and spices and mix in the fruit. Turn into a 20cm cake tin. Bake for 4 hrs at 150°C/mark 2.

Weight loss: 5%

203 Iced rich fruit cake

1680g fruit cake, rich   *Royal icing:*
70g apricot jam   300g icing sugar
410g marzipan   1 egg white
                            1 tsp lemon juice

Make the cake as in Rich Fruit Cake (No. 201) recipe. When cold spread with a thin layer of apricot jam and cover with marzipan. Make the icing by beating the egg whites and icing sugar; finally add the lemon juice.

Weight loss: 0%

204　Wholemeal fruit cake

200g margarine
200g brown sugar
3 eggs
200g plain flour
2 tsp baking powder

4g mixed spice
200g wholemeal flour
200g mixed fruit
100ml milk

Cream fat and sugar, beat in eggs. Sift white flour, baking powder and spice, add creamed mixture together with wholemeal flour and fruit. Add milk until soft. Bake at 180°C/mark 4 for 11/2-2 hrs.

Weight loss: 5%

205　Fresh cream gateau

50% sponge without fat
35% whipping cream
15% sugar

Proportions are derived from recipes and shop bought samples.

206　Gingerbread

300g flour
100g margarine
100g sugar
200g treacle

2 eggs
2 tsp ground ginger
1/2 tsp bicarbonate of soda
75ml milk

Melt the margarine, sugar and treacle in a pan, heating gently. Beat the egg well. Mix all the ingredients together and bake for about 11/4 hrs at 180°C/mark 4.

Weight loss: 11.9%

208　Lardy cake

14g fresh yeast
56g caster sugar
280ml milk
450g plain flour

5g salt
112g lard
5g mixed spice

Prepare yeast with milk. Sift flour and salt, rub in fat, add yeast and remaining milk to form a dough. Knead for 10 mins. Shape and leave for 1-11/2 hrs. Roll to 1/4in. thickness, sprinkle with sugar and spice, fold into 3, repeat twice. Shape dough and cook for 40-50 mins. at 180°C/mark 4.

Weight loss: 15%

210　Rock cakes

200g flour
100g margarine
100g sugar
3 tsp baking powder

1 egg
50ml milk
100g currants

Sift together the flour and baking powder and rub in the fat, add the currants. Mix to a soft dropping consistency with the egg and milk. Bake for about 20 mins at 190°C/mark 5.

Weight loss: 9.6%

211 Sponge cake with fat

150g flour
1 tsp baking powder
150g margarine

150g caster sugar
3 eggs

Cream the fat and sugar until light and fluffy. Add the beaten egg a little at a time and beat well. Fold in the sifted flour and baking powder. Bake for about 20 mins at 190°C/mark 5.

Weight loss: 12.9%

212 Sponge cake, fatless

4 eggs
100g caster sugar
100g flour

Whisk the eggs and sugar in a basin over hot water until stiff. Fold in flour. Bake for 25 mins at 190°C/mark 5.

Weight loss: 13.8%

214 Sponge cake with butter icing

70% sponge cake with fat
15% margarine
15% sugar

Proportions are derived from recipe review.

216 Swiss roll

3 eggs
84g sugar
70g self raising flour

15ml warm water
84g jam

Whisk eggs and sugar in a basin over hot water for until thick and creamy. Remove and continue whisking until cold. Sift baking powder, flour and salt and fold into egg mixture. Bake for 10 mins at 220°C/mark 7. Turn out, trim and spread with jam. Roll.

Weight loss: 13.8%

218 Welsh cakes

224g flour
1 tsp baking powder
pinch salt
56g butter
56g lard
112g caster sugar

1g nutmeg
0.5g cinnamon
84g currants
1 egg
56ml milk

Sift dry ingredients and rub in fat. Add sugar and fruit, mix with eggs. Roll out and cut into rounds. Cook on a griddle until brown.

Weight loss: 11.5%

219 Cheese pastry

60g Cheddar cheese
140g shortcrust pastry

Proportions are derived from a recipe review.

Weight loss: 12%

220/221 Choux pastry

100g flour
50g margarine
150ml water

2 eggs
1/4 tsp salt

Boil the water, salt and margarine, add the flour and beat over heat to form a ball of smooth mixture. Cool and beat in the eggs. Pipe out as desired and bake for about 30 mins at 200°C/mark 6.

Weight loss: 35%

222/223 Flaky pastry

200g flour
75g margarine
75g lard

1/2 tsp salt
85ml water
10ml lemon juice

Divide fat into 4 portions. Sift flour and salt, rub in one portion of fat. Mix with water and lemon juice, then knead until smooth and leave for 15 mins. Roll out, dot two-thirds with another fat portion and fold into 3. Roll out and repeat process with remaining 2 fat portions. Bake at 220°C/mark 7.

Weight loss: 24.3%

225/226 Shortcrust pastry

200g flour
50g margarine
50g lard

1/2 tsp salt
30ml water

Rub the fat into the flour, mix to a stiff dough with the water, roll out and bake at 200°C/mark 6.

Weight loss: 13.8%

228/229 Wholemeal pastry

200g wholemeal flour
50g margarine
50g lard

2g salt
30ml water

Rub the fat into the flour, mix to a stiff dough with the water, roll out and bake at 220°C/mark 7.

Weight loss: 13.6%

232 Chelsea buns

200g strong flour　　　　　35g eggs
85g skimmed milk　　　　　15g yeast
65g unfortified margarine　　55g currants
45g sugar　　　　　　　　　2g salt

Recipe from FMBRA.

Weight loss: 15%

233 Choux buns

55% cooked choux pastry (unfortified margarine)
45% double cream

Proportions are derived from dissection of shop bought samples.

234 Cream horns

300g cooked flaky pastry (unfortified margarine)
190g whipping cream
35g jam

Recipe from FMBRA.

238 Custard tart, large

250g raw shortcrust pastry　　1 egg
250ml milk　　　　　　　　　15g sugar

Make the pastry and line a shallow tin. Make a custard and use as filling. Bake until custard sets (about 40 mins) at 190°C/mark 5.

Weight loss: 11.5%

241 Doughnut with custard filling

83% doughnut
17% custard

Proportions are derived from dissection of shop bought samples.

244 Iced doughnuts

76% doughnut
12% chocolate icing
12% white icing

Proportions are derived from dissection of shop bought samples.

245 Eccles cakes

200g strong flour　　　　　55g butter
100g water　　　　　　　　75g sugar
180g unfortified margarine　370g currants
20g cooking fat

Recipe from FMBRA.

Weight loss: 19%

246 Eclairs

200g cooked choux pastry
(unfortified margarine)
150g double cream

Icing:
100g icing sugar
50g plain chocolate
30ml water

Make the choux pastry into eclairs. Slit, fill with whipped cream and top with chocolate icing.

249 Gulab jamen/jambu (homemade)

Dough:
80g full cream milk powder
80g self raising flour
40g semolina
80g fresh milk
(40g vegetable oil)

Sugar syrup:
180g sugar
230g water

Mix milk powder, flour, semolina and milk to a dough. Form into 20 small balls and deep fry until golden brown. Immerse separately in pre-prepared sugar syrup made from sugar and water boiled fiercely for 2-3 mins.

Weight loss: Dough 15%, syrup 30%

250 Gulab jamen/jambu (retail)

Dough:
720g full cream milk powder
180g semolina
40g ground almonds
20g egg
320g milk
(180g vegetable oil)

Sugar syrup:
900g sugar
1200g water

Mix milk powder, semolina, almonds, egg and milk to a dough. Form into small balls and deep fry in oil until golden brown. Immerse separately in sugar syrup as in previous recipe.

Weight loss: Dough 15%, syrup 30%

253 Hot cross buns

450g strong flour
28g fresh yeast
1 egg
pinch salt
56g margarine
112g currants
1g cinnamon
1g nutmeg
2g mixed spice

45g peel
150ml milk
60ml water
56g caster sugar

Glaze:
45g sugar
30ml milk
30ml water

Cream yeast with milk and add salt. Add to flour and eggs, and mix. Knead for 10 mins. Sprinkle with sugar, dot with fat and leave for 30 mins. Mix fat, sugar and fruit into mixture and mould. Cut a cross on each bun, glaze and bake for 15 mins at 250°C/mark 9.

Weight loss: 15%

254 Jam tarts

200g raw shortcrust pastry   200g jam

Line about 10 tart tins with thinly rolled pastry. Fill each tart with jam and bake in a hot oven, 200°C/mark 6 for 10-15 mins.

Weight loss: 6.5%

256 Wholemeal jam tarts

200g wholemeal pastry   200g jam

Method as plain jam tarts (No. 254).

Weight loss: 6.5%

257 Jellabi

Dough: *Sugar syrup:*
100g plain flour   65g sugar
20g ground rice   90g water
95g water
(35g vegetable oil)

Mix flour and rice with water to form a batter and force through an icing bag to form spirals in hot oil. Deep fry until golden brown. Immerse in sugar syrup as for Gulab jamen (homemade) (No. 249).

Weight loss: Dough 35%, glaze 30%

258 Mince pies, individual

300g raw shortcrust pastry
200g mincemeat

Roll out the pastry and cut into rounds. Place half of the rounds in tart tins. Fill with mincemeat and cover with remaining pastry. Bake for about 20 mins at 190°C/mark 5.

Weight loss: 12.6%

259 Mincemeat tart

225g raw shortcrust pastry
275g mincemeat

Line a 20in. (50cm) flan ring with pastry. Fill with mincemeat and bake for 10-15 mins at 220°C/mark 6 to set pastry, then for 20 mins at 180°C/mark 4 to cook the filling.

Weight loss: 8.0%

260 Plain muffins

300g strong flour (white)
1 tsp salt
25g margarine
225ml milk

15g fresh yeast
1 egg
flour for dusting

Sift the flour and salt, then rub in the fat. Warm the milk, blend in the yeast, add beaten egg and stir into the flour until it forms a soft dough. Beat and leave for 1-2 hrs, beat again. Roll out approximately 1cm thick and cut. Leave for 45 mins then cook on a greased griddle for 8 mins.

Weight loss: 12.3%

261 Bran muffins

150g plain flour
1 tsp baking powder
1/2 tsp salt
50g sugar
100g wheat bran
1 tsp bicarbonate of soda

200ml milk
25g butter
30ml golden syrup
1 egg
fat for greasing

Sift flour, baking powder and salt, add sugar and bran. Dissolve the bicarbonate in milk. Melt the butter and sugar together, add to dry ingredients with milk and egg, and mix. Spoon into tins and bake at 200°C/mark 6 for 15-20 mins.

Weight loss: 16%

262 Pinni

120g brown flour
180g sugar

50g almonds
120g butter ghee

Fry flour slowly, add sugar and chopped nuts.

Weight loss: 0%

263 Rum baba

Cake:
112g flour
pinch of salt
7g yeast
75ml water
1 egg
14g sugar
42g unfortified margarine
28g almonds

Rum syrup:
45ml rum
300ml water
112g sugar
20ml lemon juice

Warm flour and sieve with salt. Cream yeast with half the water, add to flour mixture and leave in a warm place for about 15 mins. Add beaten egg and remaining water, beat well. Add sugar and fat, leave until doubled in size. Beat again, fill rum baba moulds one third full and store until contents have doubled in size. Bake at 220°C/mark 7 for 15 mins. Turn out and soak in hot syrup.

Weight loss: 15% for cake, 15.2% for syrup

264 **Cheese scones**

200g flour 50g margarine
14g baking powder 60g cheese
1g salt 125g milk

Sift dry ingredients and rub in fat. Stir in grated cheese and mix in a bowl. Roll out and cut into rounds. Bake in a hot oven at 220°C/mark 7 for 10 mins.

Weight loss: 15%

266 **Plain scones**

200g flour 50g margarine
4 tsp baking powder 10g sugar
1/4 tsp salt 125ml milk

Sift the flour, sugar and baking powder and rub in fat. Mix in the milk. Roll out and cut into rounds. Bake in a hot oven at 220°C/mark 7 for about 10 mins.

Weight loss: 18.5%

267 **Potato scones**

112g plain flour 56g margarine
112g mashed potato 14g baking powder
1g salt 65ml milk

Sift flour and salt, rub in fat and add potato. Mix in raising agent, add milk to form a soft dough, knead until smooth. Roll out until 1.5-2cm thick and cut into rounds. Brush tops with milk, bake for about 7-10 mins. at 220°C/mark 7.

Weight loss: 10.3%

268 **Wholemeal scones**

200g wholemeal flour 50g margarine
14g baking powder 10g sugar
1g salt 125ml milk

Method as recipe for plain scones (No. 266).

Weight loss: 14%

269 **Wholemeal fruit scones**

200g wholemeal flour 125ml milk
14g baking powder 10g sugar
1g salt 50g sultanas
50g margarine

Method as recipe for plain scones (No. 266). Add fruit to the dry ingredients with sugar.

Weight loss: 14%

270 Scotch pancakes

200g flour
1/2 tsp salt
1 tsp cream of tartar
50g margarine
1/2 tsp bicarbonate of soda

25g caster sugar
1 egg
200ml milk

Sift flour with salt and raising agent, rub in fat and mix in sugar. Add egg and milk to give a stiff batter. Cook by spoonsful on hot greased griddle.

Weight loss: 9.4%

271 Sevyiaan

215g vermicelli
200g butter
300g milk
100g sugar

50g water
20g sultanas
5g dried coconut
15g almonds

Fry vermicelli in butter for one minute until brown. Add milk, sugar, water, sultanas and coconut and cook over a gentle heat, stirring until mixture thickens. Pour into dish and decorate with almonds.

Weight loss: 24.9%

272 Strawberry tartlets

Pastry: 145g cooked shortcrust pastry (unfortified margarine) shaped into 12 cases

Fruit: 12 fresh strawberries (144g)

Glaze: 1 tsp arrowroot   20g sugar
125ml water   2-3 drops lemon juice

Fill pastry cases with fresh strawberries and cover with glaze.

Weight loss: Pastry 13.8%, glaze 12.7%

275 Vanilla slices

270g cooked flaky pastry (unfortified margarine)

Filling:
30g sugar
50g eggs
205g skimmed milk
10g cornflour

Icing:
35g jam
205g sugar
14g water

Recipe from FMBRA.

Weight loss: For pastry and custard filling, 10%.

276 Waffles

84g butter
2 eggs
375ml milk

280g self-raising flour
1/2 tsp salt
1 tsp baking powder

Sift dry ingredients, then add egg yolks, cooled melted butter and some of the milk. Beat in remaining milk. Whisk egg whites and fold into batter. Cook on a waffle iron until steaming stops.

Weight loss: 32%

277 Welsh cheesecake

200g raw shortcrust pastry
60g raspberry jam
60g eggs
60g margarine

60g sugar
60g flour
4g baking powder

Prepare as for Bakewell tart (No. 283), but using flour instead of ground almonds.

Weight loss: 6%

278/280 Apple pie, one crust, plain or wholemeal

200g raw plain or wholemeal shortcrust pastry
450g cooking apples, prepared
80g sugar

Method as for fruit pie, one crust (No. 308/311).

Weight loss: 4.2%

279/281 Apple pie, two crust, plain or wholemeal

450g raw plain or wholemeal shortcrust pastry
450g cooking apples, prepared
80g sugar

Method as for fruit pie, two crust (No. 309/312).

Weight loss: 4.2%

283 Bakewell tart

200g raw shortcrust pastry
60g raspberry jam
60g eggs.
60g margarine

60g sugar
10g flour
60g ground almonds

Line a 17in. (43cm) flan ring with pastry. Cover base with jam. Cream fat and sugar, add beaten egg gradually and beat well. Fold in almonds and flour. Turn into flan case and bake at 190°C/mark 5 for 30 mins.

Weight loss: 6.0%

284/285 Blackcurrant pie, two crust, plain or wholemeal

450g raw plain or wholemeal shortcrust pastry
450g blackcurrants
80g sugar

Method as for fruit pie, two crust (Nos. 309/312).

Weight loss: 4.2%

286 Bread and butter pudding

60g bread
15g butter
500ml milk

25g sugar
2 eggs
25g currants

Cut bread thinly and spread with butter. Beat eggs with sugar and add milk. Layer bread and currants in a dish and pour the eggs and milk over bread. Leave to soak for 30 mins, then bake at 180°C/mark 4 for 30-40 mins.

Weight loss: 23.5%

287 Bread pudding

225g bread
275ml milk
50g melted butter
75g demarara sugar

4g mixed spice
1 beaten egg
175g dried fruit

Break bread into pieces, cover with milk and leave for 30 mins. Add remaining ingredients, mix well and bake for 1 1/4 hrs at 180°C/mark 4.

Weight loss: 24%

288 Cheesecake

Base for 18cm tin:

150g digestive biscuits
75g margarine

Top:
350g cream/curd cheese
25g cornflour
140g double cream
1/2 tsp vanilla essence

2 eggs
100g caster sugar
1 lemon (40g juice and finely grated rind)

Melt the margarine in a pan and combine with the biscuit crumbs. Press into the base of the tin. Combine the topping ingredients and pour into base. Bake for 45 mins at 180°C/mark 4, until only just firm in the centre.

Weight loss: 5.6%

290 Christmas pudding

100g flour
300g fresh breadcrumbs
1 tsp mixed spice
1/2 tsp salt
125g suet
150g raisins
150g sultanas

150g currants
50g chopped mixed peel
30g ground almonds
150g brown sugar
3 eggs
15g treacle
150ml stout

Sift the flour, spices and salt into a basin and mix in all dry ingredients. Whisk the eggs, treacle and stout and stir thoroughly into dry ingredients. Put into well greased basins, cover with greased paper and foil. Boil for 6 hrs. Renew foil and store. Reboil for 2 hrs before serving.

Weight loss: 0%

292 Crumble with pie filling

400g commercial fruit pie filling 100g flour
50g margarine 50g sugar

Rub in flour, margarine and sugar. Arrange over filling in pie dish. Bake for 40 mins at 190°C/mark 5.

Weight loss: 4.3%

293 Apple crumble

400g cooking apples (weighed 50g margarine
after preparation) 100g sugar
1/2 tsp cinnamon 100g flour

Peel, core and slice apples. Arrange in a dish and sprinkle with sugar. Rub together the other ingredients and pile on top. Bake for 40 mins at 190°C/mark 5.

Weight loss: 7.4%

294/295 Fruit crumble, plain or wholemeal

400g prepared fruit 100g plain or wholemeal flour
50g margarine 100g sugar

Method as for apple crumble (No. 293).

Weight loss: 7.4%

296/297 Custard

500ml whole or skimmed milk
25g custard powder
25g sugar

Blend the custard powder with a little of the milk. Add sugar to the remainder of the milk and bring to the boil. Pour immediately over the paste, stirring all the time. Return to the pan, bring back to boiling point, stirring.

Weight loss: 20.9%

299 Confectioners' custard

1 egg 14g plain flour
28g caster sugar 150ml milk
2-3 drops vanilla essence

Cream the sugar and egg, then fold in flour. Stir in milk and boil while stirring for 2-3 mins.

Weight loss: 20.6%

301/302 Dream Topping made up with milk

40g pkt Dream Topping powder
150 ml 300ml whole or skimmed milk

Sprinkle powder onto milk. Whisk for 2 mins.

303 Eve's pudding

500g sponge cake
500g fruit (e.g., apple, gooseberry, rhubarb, plum)

Proportions are derived from dissection of samples.

304 Fruit flan, pastry base

98g raw flan pastry
425g tinned fruit

Glaze:
15g cornflour
150ml fruit juice

Bake pastry, fill with fruit and chill. Cover with glaze.

Weight loss: 13.8% for pastry, 12.7% for glaze

305 Fruit flan, sponge base

1 sponge flan base (180g)
425g tinned fruit

Glaze:
15g cornflour
150ml fruit juice

Arrange fruit in case and chill. Coat with glaze and leave to set.

Weight loss: 12.7% for glaze

306 Flan case, pastry

224g flour
1/4 tsp salt
142g margarine

42g caster sugar
2 egg yolks

Prepare as shortcrust pastry (Nos. 225/226), add sugar after margarine.

Weight loss: 13.8%

307 Flan case, sponge

2 eggs
56g caster sugar
42g flour

Method as for fatless sponge cake (No. 212).

Weight loss: 13.8%

308/311 Fruit pie; one crust, plain or wholemeal

200g raw plain or wholemeal shortcrust pastry
450g prepared fruit
80g sugar

Place fruit in a pie dish and cover with pastry. Bake for 10-15 mins at 200°C/mark 6 to set pastry, then about 20 mins at 180°C/mark 4 to cook fruit.

Weight loss: 4.2%

309/312 Fruit pie, two crust, plain or wholemeal

450g raw plain or wholemeal shortcrust pastry
450g fruit (e.g., apple, gooseberry, rhubarb, plum)
80g sugar

Line a pie dish with half the pastry. Fill with prepared fruit and sugar and cover with remaining pastry. Bake for 10-15 mins at 220°C/mark 7 to set the pastry, then for about 20-30 mins at 180°C/mark 4 to cook the fruit.

Weight loss: 4.2%

314/315 Instant dessert powder - made up

66g instant dessert powder
300ml whole or skimmed milk

Sprinkle powder onto milk. Whisk briskly.

316 Lemon meringue pie

200g raw shortcrust pastry
2 lemons (80g juice)
2 eggs
125g caster sugar
25g cornflour
15g margarine
125ml water

Boil the cornflour, water, grated rind, lemon juice and 25g of sugar. Cool, stir in egg yolks, and pour mixture into the pastry case. Make a meringue with egg whites and rest of sugar; pile on top of the lemon mixture. Bake for 30 mins at 180°C/mark 4 until crisp and brown on top.

Weight loss: 19.0%

317/318 Milk puddings

500ml whole or skimmed milk
50g cereal (rice, sago, semolina, tapioca)
25g sugar

Simmer until cooked or bake in a moderate oven 180°C/mark 4, according to type of cereal.

Weight loss: 19.1%

319/320 Sweet pancakes

100g flour
250ml whole or skimmed milk
1 egg
50g lard
50g sugar

Sieve the flour into a basin, add the egg and about 100ml of the milk, stirring until smooth. Add the rest of the milk and beat to a smooth batter. Heat a little lard in a frying pan and pour in enough batter to cover the bottom. Cook both sides and turn onto sugared paper. Dredge lightly with sugar. Repeat until batter is used, to give about 10 pancakes.

Weight loss: 20%

321/322 **Pie with pie filling, two crust, plain or wholemeal**

450g raw plain or wholemeal shortcrust pastry
450g fruit pie filling

Method as for fruit pie, two crust (Nos. 309/312).

Weight loss: 4.2%

323 **Queen of puddings**

250ml milk	2 separated eggs
25g butter	rind of 1 lemon
50g fresh breadcrumbs	50g jam
100g sugar	

Heat the milk and pour over the breadcrumbs and 30g sugar. Leave to soak for 30 mins. Add beaten egg yolks and grated lemon rind and pour into greased pie dish. Bake for 20 mins at 180°C/mark 4. Spread the top with jam. Whisk the egg whites until stiff, then whisk in the rest of the sugar one teaspoonful at a time. Pile on top and bake at 140°C/mark 1 until crisp golden brown.

Weight loss: 8.5%

325 **Steamed sponge pudding**

100g flour	50g caster sugar
1 tsp baking powder	1 egg
50g margarine	30ml milk

Cream the fat and sugar. Beat in the eggs a little at a time. Fold in the sifted flour and baking powder, adding milk to give a soft dropping consistency. Turn the mixture into a greased basin and steam for about 21/2 hrs.

Weight gain: + 3.9%

326 **Sponge pudding with dried fruit**

88% steamed sponge pudding
4% currants
4% sultanas
4% raisins

Method as steamed sponge pudding (No. 325). Proportions derived from a recipe review.

Weight loss: 0%

327 **Sponge pudding with syrup or jam**

88% steamed sponge pudding
6% jam
6% golden syrup

Method as steamed sponge pudding (No. 325). Proportions are derived from a recipe review.

Weight loss: 0%

329 Spotted dick

50g flour
50g breadcrumbs
50g shredded suet
30g sugar
25g currants

1 tsp baking powder
1/4 tsp salt
80ml milk

Method as suet pudding (No. 330) with currants added to dry ingredients.

Weight loss: 0%

330 Suet pudding

50g flour
50g breadcrumbs
50g shredded suet
30g sugar

1 tsp baking powder
1/4 tsp salt
80ml milk

Mix the dry ingredients to a soft paste with the milk. Pour into a greased basin, cover with greased paper and steam for about 2 1/2 hrs.

Weight loss: 0%

331 Treacle tart

300g raw shortcrust pastry
250g golden syrup
50g fresh breadcrumbs

Line shallow tins with pastry, pour in the syrup and sprinkle with the breadcrumbs. Bake for 20-30 mins at 200°C/mark 6.

Weight loss: 0%

332 Trifle

75g sponge cake
25g jam
50g fruit juice
75g tinned fruit
25ml sherry

250g custard
25g double cream
10g nuts
10g angelica and cherries

Slit the sponge cake, spread with jam and sandwich together. Cut into 4 cm cubes. Soak in the fruit juice and sherry. Mix with the fruit, cover with cold custard and decorate with the whipped cream, nuts and angelica.

334 Trifle with Dream Topping

As trifle (No. 332) except substitute 25g of made up Dream Topping for 25g double cream.

336 Chevda/chevra/chewra

17g dried lentils
16g water
35g fresh peanuts
18g dried rice flakes

8g vegetable oil
2g mixed spices
2g chilli powder
2g salt

Soak the lentils in water overnight, drain and dry. Fry lentils, peanuts and rice separately in vegetable oil. Mix the cooked ingredients with the spices, chilli powder and salt.

Weight loss: 19.5%

340 Dumplings

100g flour
45g suet
75g water

1 tsp baking powder
1/2 tsp salt

Mix the dry ingredients together with the cold water to form a dough. Divide into balls, flour them and place in boiling water. Boil for 30 mins.

Weight gain: + 52.7%

342 Macaroni cheese

280g cooked macaroni
350ml milk
25g margarine

25g flour
100g grated cheese
1/2 tsp salt

Boil the macaroni and drain well. Make a white sauce from the margarine, flour and milk. Add 75g of the cheese and season. Add the macaroni and put in a pie dish. Sprinkle with remaining cheese and brown under the grill or in a hot oven at 220°C/mark 7.

Weight loss: 9.4%

345 Pakoras/bhajia

150g aubergine
100g potatoes
75g spinach
30g onion
150g chick pea flour

4g salt
5g ground dried chillies
1g crushed coriander seeds
150g water
(30g vegetable oil)

Slice vegetables thinly. Make a batter with flour, salt, chilli powder, coriander seeds and water. Dip vegetables in batter and deep fry in oil until golden brown.

Weight loss: 28%

346/347 Savoury pancakes

112g flour
300ml whole or skimmed milk
1 egg

56g lard
1/4 tsp salt

Method as for sweet pancakes (No. 319/320).

Weight loss: 20%

348 Pilau rice

450g Basmati rice
168g ghee
1 chopped garlic clove
10 cloves
10 cardamoms

112g peeled and chopped onion
1 tsp salt
112g sultanas
56g blanched almonds
1800ml water

Heat oil, fry onion until soft, add garlic, cardamoms and cloves and continue frying for 2 mins. Add rice and boiling water, bring back to the boil and then simmer until all the water is absorbed. Fry sultanas and almonds in a little oil and add to rice.

Weight loss: 35.6%

349 Cheese and tomato pizza

Dough:
200g flour
1 tsp salt
1 tsp sugar
150ml warm water
15g fresh yeast or 2 tsp dried yeast

Topping:
200g tomatoes
150g cheese
8 black olives (40g)
20g oil

Make the dough, proving once. Knead and roll out shape. Leave for 10 mins. Arrange sliced or pulped tomatoes on top, then cheese and olives. Brush with oil. Bake for 30 mins at 230°C/mark 8.

Weight loss: 14.1%

352 Risotto

224g long grain rice
550g stock
84g chopped onion

56g margarine
1 tsp salt
1g pepper

Melt margarine, add onion and fry until soft. Add washed rice and stir over low heat for 10 mins. Pour in stock, bring to the boil and simmer until all is absorbed.

Weight loss: 37%

353 Meat samosas

80g flour
10g butter
25ml water
5g oil (for frying ingredients)
70g minced lamb

10g fresh green chillies
15g chopped onion
30g chopped potatoes
30g peas
(190g vegetable oil)

Make a dough with flour, butter and water, divide into two rounds and roll out. Sandwich these together with a little oil and cook in a frying pan without fat for 1 min on each side. Divide into 4 quarters and separate to give 8 samosa cases. Fry minced lamb in oil with chillies, onion, potato and peas. Fill the cases with the mixture, fold over pastry edges to make a triangle, seal with flour and water paste. Deep fry in oil.

Weight loss: 13.5%

354 Vegetable samosas

80g flour
10g butter
25ml water
25ml oil (for frying ingredients)
1/2 tsp salt
15ml lemon juice

260g boiled potatoes
45g chopped onion
63g pea
4.5g mixed spice
15ml water
(190g vegetable oil)

Make cases from flour, butter and water as for meat samosas (No. 353). Fry vegetables, lemon juice and spices in oil for 5 mins. Add water and cook for a further 5 mins. Fill the cases with the mixture, fold over pastry edges to make the triangle, seal with flour and water paste. Deep fry in oil.

Weight loss: 25.9% for filling, 13.5% for pastry

355 Sev/ganthia

220g chick pea flour
4g salt

105g water
(55g vegetable oil)

Mix flour, salt and water to make a thick batter. Force through a sev/siawa machine into hot oil and deep fry until golden brown.

Weight loss: 30.4%

358 Sage and onion stuffing

224g onion
112g white breadcrumbs
4g sage
1g salt

1/2g pepper
1 egg
56g margarine

Slice onions and sage, parboil, drain and chop, mix with breadcrumbs. Melt margarine and add to stuffing. Mix thoroughly.

Weight loss: 19.3%

359/360 Yorkshire pudding

100g flour
1 tsp salt
1 egg

250ml whole or skimmed milk
20g dripping

Sieve flour and salt into a basin. Break in the egg and add about 100ml of milk, stirring until smooth. Add the rest of the milk and beat to a smooth batter. Pour into a tin containing the hot dripping. Bake for about 40 mins at 220°C/mark 7.

Weight loss: 16%

FOOD INDEX

- Foods are indexed by their food number and **not** by page number.

- Numbers in brackets refer to foods where limited nutritional information is given in footnotes. Other aspects of the composition may differ from the values given in the main tables.